A QUEST FOR

BEES

— IN BHUTAN —

Best wishes,

Pamela Carmen

In 'A Quest for Bees in Bhutan', Paula Carnell elegantly reveals her burgeoning and transformative connection with bees... her down to earth narration serves to subtly underscore the unique value of bees and beekeeping to the natural environment and the revitalisation of the human spirit.

———

Two definitions of democracy:

1. The common people, considered as the primary source of political power.
2. Majority rule.

Gross National Happiness is the best definition for Democracy I've heard. I love learning when I read.

A QUEST FOR
BEES
— IN BHUTAN —

BY
PAULA CARNELL

Published by Paula Carnell
www.paulacarnell.com
Email: paula@paulacarnell.com

First published by Paula Carnell
September 2019
Castle Cary, Somerset BA7 7JW.

Created by Paula Carnell
Copyright © Paula Carnell 2019
Inner photograph of Paula by Janette Edmonds
www.beautifully-you.photography
Graphic design Andy Batchelor
Photography by © Paula Carnell

Publisher Paula Carnell

ISBN: 978-0-9562073-3-3

FOR WANGCHUK, SONAM, DAMCHOE, NAVIN, PRITI,
KABIRAJ, ALL IN THE BHUTAN BEEKEEPERS COOPERATIVE
PEMA C AND PEMA D. THANK YOU FOR MAKING MY VISIT
TO YOUR KINGDOM SO SPECIAL, AND FOR THE DIVINE SPIRIT
THAT DREW ME TO YOU. MAY WE ALL MEET AGAIN SOON.

A donation of £1 from the sale of each copy of this book will go to
'Opening your heart to Bhutan', Registered Charity no. 1165794
www.openingyourhearttobhutan.com

'If I was still using a wheelchair, my trip to Bhutan
would not have been possible'.

Contents

INTRODUCTION

A trip to Bhutan was intended to give me insight and material for a single chapter in my next book, 'Spiritual Bee-ing', however the people I met, the landscape, temples and of course the bees, decided that there was more than a single chapter to share about this magical Kingdom of the Thunder Dragon. By reading my tales, I hope that you too capture a glimpse into this last sanctuary on our beautiful planet.

Paula J Carnell

Bhutan

Bumdrak

Thimphu

PARO

Dochula

Run...

Dagana

- Town
- ✈ Airport
- Road
- River
- ⛰ Mountains
- ---- Trek

1. WHY BHUTAN?

HAT IF THERE WAS A PLACE ON EARTH WHERE THE bees were happy and healthy? Acres of forest and wildflower meadows feeding bees and other birds and insects. A life where humans and nature live in harmony together. I'd heard about a small Himalayan kingdom, and had friends who'd visited, returning with tales of a tranquil and beautiful place, millions of miles away from our modern ways of living. Bhutan was calling to me as a perfect place to celebrate turning fifty. 'Gross National Happiness' sounded like a perfect measure of a country's success. Bhutan has been making decisions on its development using its gross national happiness (GNH) guidelines as an indicator since 1998. GNH is an attempt to live in a way that's "holistic," not restricted to merely measuring economics like the gross domestic product, or GDP. In the film 'The Other Final', the Minister of Foreign Affairs in Bhutan, Lyonpo Jigme Y Thinley, states that " the realisation that every human being searches for happiness in life, and if that is the ultimate goal, of every Bhutanese, every human being, then it becomes the responsibility of the Government". Surely a country with ministers who believe this is somewhere that deserves more understanding, and observation. As we in the West are starting to question the viability of Western liberal capitalism, the idea of a kingdom in tune with nature appealed to me greatly. Many of us are searching for a new way of living and doing business, no longer at the cost of the earth on which we live. So I began on a quest to the top of the world in search of not only happy people, but happy bees. With hope that I could share the wisdom and experiences of the Bhutanese back at home in the West, I had no idea then about the impact such a trip would have on me.

Everest - the top of the world.

2. PARO

EAVING THE SMOGGY POST DIWALI AIR OF DELHI

and flying North East towards the Himalayas, our Drukair pilot was eager to entertain his carriage. Filled with a mixture of Bhutanese, Indians and Western tourists, as we soared into the clear blue skies above the clouds, white peaks of mountains could be seen. We were directed to look to the left to the grand peaks, including the highest of them all, Everest. I had a window seat with a perfect view of the actual top of the world. Since recovering from my Ehlers Danlos Syndrome (EDS) I felt strongly that I should celebrate my fiftieth birthday literally 'on top of the world' to mirror how I felt with such unexpected, and hard gained health. As I was handed cameras from passengers not lucky enough to have my seat, my eyes welled up as we collectively shared

the wonder and majesty of such a sight. The pilot kept us steady as half the plane left their seats to grab a view of this most spectacular sight. Living moment by moment with such a busy build up prior to this trip and trusting the travel plans of my agent Maria, I hadn't looked at my route and so this view was an unexpected bonus. Soon we were entering Bhutanese air space, expertly flown by one of only eight pilots granted permission to land in this protected Kingdom. Paro airport is a short strip in the valley completely surrounded by tree coated mountains. My husband

Entering Paro airport

Greg was justifiably anxious about my journey here. My first landing was smooth and exciting as we glided past Tiger's Nest temple and the Dzong before landing at a quiet airport. Unlike other airports, we were disembarking the only plane on the runway. The passengers took their time taking selfies against the backdrop of highly decorated terminal buildings. It was surprisingly warm with the sun shining on us, so I couldn't

The Royal family.

claim it was the cold was causing my eyes to water. No buses or long passage-ways to lead us to customs. A large photograph of the Royal Family hung against the terminal building, an indication of the nation's genuine affection for their young King and his family. I was greeted after collecting my bags by a guide, unaware at that point of the custom for all tourists to be escorted at all times by a guide and a driver. It felt quite extravagant as a solo traveller, however, after driving across the country I soon realised how a guide certainly couldn't talk and drive at the same time! The twisty mountain roads took full concentration and a lifetime of experience to navigate safely. A wise rule in Bhutan is that tourists are not permitted to drive here. It had been a long and tiring journey from Somerset to London, then Oman and Delhi, before finally arriving in Paro Bhutan. Being led to a large clean room with a spectacular view, three days after leaving home, I blessed my travel agent Maria for booking me a room at The Village Lodge with a view straight across the valley facing the Tiger's Nest Monastery. This symbol of Bhutan has to be the most photographed building in the Kingdom, and I later learned that it is a compulsory part of every tourists' itinerary. After a delicious home cooked dinner of roasted vegetables and rice, I retired to my large and very comfortable bed. One last look at the Tiger's Nest in the darkness, butter lamps glowing from it and several other smaller temples above and below it. Despite the below freezing temperatures, I opened my window and listened to the silence. Night time in Bhutan isn't filled with passing aircraft, busy roads or people, but a faint sound of monks chanting and the much louder sound of barking dogs...

3. BUMTHANG

A WOKEN AT 5AM BY MY ALARM FOR A 5.30 BREAKFAST

wasn't pleasant. I hadn't slept well, waking up every hour. Unable to adjust to the extremely warm heated bedroom, opening a window to the below freezing outside air was too extreme. A bonus however was managing a good web chat with Greg at 1.30am. Breakfast was fresh fruit and I sat and waited for the guide to collect me. The sun was coming up and it was cold. A large Stupa filled the view from the dining room window. On arriving, my guide said we had plenty of time, so he had a cup of tea. I was eventually dropped off at the airport (only a 5 minute drive away) at 6.20 and a large queue was forming, which was for the international flights to India and Thailand. The queue for my internal flight to Bumthang was very much shorter. I struggled with not

feeling anxious, what if I was in the wrong queue or missed my flight. I stayed close to people I recognised who were checking into the same Bumthang flight. The airport was beautiful, small and interesting. The shops were more like Communist Russia, basic and only taking only cash payments. I found some small pots of honey, 'Puthka the antibiotic honey' in between the almost bare shelved gift shop. Awaiting a delivery explained the emptiness. Collecting four jars, enough for me and gifts for my 'bee team', I was

The airport bookshop.

Bumthang airport

pleased that the 25ml jars wouldn't cost me excess baggage, costing around £3 each, a perfect gift and as I later learned a justifiable expense. The bookshop was incredible, with Buddhist books, and any new age, mind-expanding book you can imagine, alongside guide books and stunning (heavy) photographic books on Bhutan. I bought a handy sized map with estimated timings between key locations. I also bought some postcards and beautiful assortment of stamps. Bhutan is famed for its unusual and collectable stamps. I was exhausted before the plane even came in for us to board. There was a half hour delay due to turbulent weather- it looked fine from where we were. With only 10 people on my flight, I learned later that they only take 18 due to weight limits and the high altitude, and I needn't have worried about securing a window seat. We all had one! It was quite bumpy but with dramatic views. We were able to see Bhutan's highest peak at 7538m, Kula Kangri, a sacred mountain forbidden to be climbed. The previous day's smooth flight had not prepared me for the genuine concern as we dropped through the clouds inbetween unseen mountains. The descent was bumpy but thankfully we all arrived safely, slightly bemused at the welcome. We landed and walked straight from the plane into a field where our guides, and friends and family of Bhutanese, were waiting for us. Two very small buildings housed the airport security, and a mobile fire brigade were on hand, just in case. An ancient tractor and trailer followed us from the plane with our bags ready for us to collect. This was definitely the smallest airport I had ever flown into.

Drukair runs the half hour flights across Bhutan a few days a week. The service is often cancelled or delayed due to the weather and altitude. When planning a trip, it's always advised to allow to be delayed by a few days each way. The flight saves a twisty eight to ten-hour drive. I'd made the decision to risk flying both ways. I love being in the air looking down on the landscape below, and this country definitely had some of the most spectacular aerial views. I was met by Wangchuk and Sonam, my new guide and driver who were going to get to learn a lot about bees. Wangchuk was a young 29-year-old who had changed career from helping his parents on their small farm. He would take the cattle high into the mountains near his home in Trongsa, about 3 hours west of Bumthang. From a solitary life, spending weeks travelling remote pastures and woodland, sleeping in the huts used by herdsman for centuries, he had retrained as a guide for tourists, and now shared a flat with his brother in Bhutan's capital city Thimpu. The capital is a twisty hair raising 8-hour drive back towards Paro. Sonam was older, in his mid-thirties, and was a former monk. I thought this was interesting and unusual, however in a country where until ten years ago, the eldest son of every family became a monk, this wasn't odd at all. He was now married and had two young children, but enjoyed his job of driving tourists and guides backwards and forwards along the main road of Bhutan, linking Paro in the west with Tashigang in the far east. I asked how it is regarded to leave the monastery, he said the monks say he will just live a life with more challenges. I very much felt over the coming days that I was his current challenge! Well briefed before my arrival, Wangchuk had already made arrangements for me to meet with some local beekeepers and the Beekeeping Cooperative of Bhutan. The headquarters happened to be on the same road as my hotel so we could

'We were able to see Bhutan's highest peak at 7538m, Kula Kangri, a sacred mountain forbidden to be climbed'

actually walk along the Somerset levels styled track, lined with autumnal willows. I had been booked to stay in Bumthang district for 8 nights. Again, most tourists would only be here for a day or two, long enough to visit the impressive Dzong (regional council offices) with its incorporated monastery and temple and a couple of other sacred sights. The Dzong was positioned half way up a mountain, overlooking the airport, and I could see it from my bedroom window each morning. This place became my new home.

TREGONA MELIPONA (STINGLESS BEE) Smaller in size than Apis mellifera at around 3-5mm in size and much slimmer yet with a stronger tolerance to predators and diseases due to their extensive use of propolis. They collect nectar from a 300-500metre radius of their hive and forage on native plants of tropical and sub-tropical areas at a lower altitude (700-1500 metres above sea level) and so found across southern Bhutan. As harmful pesticides and synthetic fertilisers are not permitted in Bhutan, the farmers rely heavily on this species of bee to pollinate their crops. Stingless bee 'Puthka' honey has a strong citrusy taste and a high water contented which has been proven scientifically to kill cancer cells. One colony of these solitary bees produces 500grams of honey each year in tiny wax cups. There are now ways of harvesting the honey without killing the bees, a benefit for all concerned.

The willow lined single track to my hotel in Bumthang

4. NAVIN & PRITI: APIS MELLIFERA

AFTER DROPPING OFF MY BAGS AND HAVING A SECOND

breakfast, we drove the short distance to the Bumthang Beekeepers Cooperative buildings. I had been trying to research the bees in Bhutan as much as possible before I arrived. My main source of information had been the charity, Bees for Development and in particular, Nicola Bradbear's research report following her trip in 1986. I had also met Nicola at the Dutch 'Learning from the Bees' conference in August where she had advised me about asking to see Apis cerana bees rather than asking about 'traditional beekeeping'. Bees for Development works globally helping combat poverty and biodiversity by assisting and training beekeepers. Nicola's extensive experience with various species of honey bees encouraged me to presume her advice about not introducing Apis mellifera into Bhutan was taken. No one was in at the cooperative, however there were a number of beehives positioned around the grounds. I was interested to see bees flying despite the cold, and that they looked very much like my own bees at home. I hadn't before seen Apis cerana bees so presumed that was what they were. All the hives were Langstroth style, a brood box covered with a roof and a large stone, carried up from the river bank. I had heard about beekeepers training in India and introducing modern beekeeping methods to Bhutan. I was eager to meet up with a beekeeper and learn more about the situation there. Opposite the Cooperative site my guide had found a chicken farmer who agreed to meet with me. Navin, and his wife Pritti, were southern Bhutanese who were also honey farmers. They have been keeping bees for eighteen years and he currently has 150 hives.

Honey farmers Navin and his wife Pritti.

The Langstroth style hices, a brood box all covered with a roof and a large stone

It had been difficult for me to find out about the climate or honey harvest times before I arrived. The honey harvest in Bhutan was reported as July but also November and December. I had caught Bumthang at the post-harvest, Winter preparation time. The hives were being reduced to a single brood box after an autumnal feed. Navin and Pritti invited me into their home where I was told more about their honey crops, their bees and their dreams. I always love to know what people's dreams are. Although not always in a position to help directly, I store those dreams in my memory, ready to share with those who can help realise them. Navin had left the Beekeeping cooperative as the payments for honey were too low. He was producing 5 metric tons of honey per year but didn't have the market to sell it to. The Bhutanese are mostly Buddhists and to my surprise, they believe that eating honey is a sin. Buddha had apparently been punished for 100 years for taking just a single drop of honey on his tongue. My poor driver Sonam, was already being challenged within his first few hours with me, we were tasting honey, and not just single drops!

Navin had some excellent buckwheat honey and was experimenting with single varieties, including orange, eucalyptus and rosewood. With the majority of Bhutanese not eating honey, the market was restricted to European and American settlers, Indians or the tourist market. The main issue with tourists was the weight restrictions on flights. Internal flights had a maximum of 15 kg meaning most visitors just couldn't add heavy honey as souvenirs. Navin also explained to me that the bees he had were in fact Apis mellifera. I was absolutely stunned. It hadn't occurred to me that anyone would have introduced this species. Navin continued to tell me how, in the late 1980s a Swiss Business man, Fritz, who has since remained in Bhutan, introduced 5 colonies of Apis mellifera bees to the beekeeping cooperative. He brought modern methods of beekeeping and the education of pollination to the Bhutanese used to small scale honey production in traditional and log hives for bees. After many cups of tea and a jar of his finest honey, Navin offered to help introduce me to the Cooperative Beekeepers and answer any more questions I may have during my stay. I fell in love with Priti and her children as she spoke of her passion to run a successful business. I knew that we'd be keeping in touch and I really want to find a way to help them. As the weekend approached, I needed patience before more bee work could be organised.

APIS MELLIFERA *The western or European honey bee was imported to Bhutan from India after its widespread distribution from Europe for honey production and farming. Although drones are said to chase Apis cerana queens, so far the species hasn't adapted to the wild in Bhutan or cross bred with the native Apis cerana bees. They are kept in managed wooden hives using frames holding wax foundation.*

5. TEMPLES & PEMA

HE NEXT TWO DAYS WERE
SPENT EXPLORING THE

sacred places of Bumthang. Wangchuk had planned for me to visit the Beekeeping Cooperative member Kabiraj who happened to live a short drive, less than 2 minutes, from my hotel. This wasn't until Sunday so I could relax and do the normal tourist things. Jakar Dzong overlooks the Chokhor valley and was chosen as a site in 1549. When the monks gathered to select a site for a monastery, a large white bird appeared and settled on the spur of a hill. This was considered an important omen and was then named 'Castle of the White Bird'. Lama Ngagi Wanchuk was the founder of the monastery. As a Tibetan monk, he became the founder of Bhutan, uniting all the tribes into one kingdom. There were impressive views from this dramatic and ancient building. Dzongs not only house the council offices, but also a monastery and temples. We could enter the temples after removing shoes and hats and leaving cameras outside. Phones are permitted, (but not to take photographs) and throughout my travels I continued to be amused by monks on their mobile phones. There was a small temple above the Dzong where the naming of babies occurs. During my trip I noticed that many people appeared to have the same name, and there wasn't a clear difference between female or male names. I learned that during naming ceremonies, a monk picks a name out of a bowl. This explains somewhat, the limited selection. We then visited Jampey Lhakhang, a really ancient temple built in 659 by the Tibetan King Songsten Gampo. There are three stone steps inside the main temple which signify the three ages, past, present and future Buddha. The inner temple is dedicated to Jampa, the future Buddha who has his feet on an elephant. This is the oldest part of the oldest chapel in Bhutan and I was incredibly lucky that Sonam knew the monk on watch and so he unlocked this sacred chamber and we were able to go inside and offer blessings to this impressive ancient statue. The steps were sinking into the ground, leaving only half of the middle step, implying that we were in the time of the present Buddha. When the second step had fully sunk it would be time for the future Buddha to rise. It is believed that the current Buddha is defeating the evil forces in the West, however the evil is returning to

Wangchuk & Sonam

the East and so the Bhutanese are preparing for a battle in the near future against the evil forces. When the third step sinks the world ends... There was time for one more temple to visit, Kurjay Lhakhang. Yak herder women were sat outside with large blankets filled with ancient looking items- meditation bowls, bells and candle holders along with beautiful woven wool and silk scarves. I was conscious that my two male guides weren't so interested in this first opportunity for shopping, so made a mental note of items and prices so we could return for a more focussed shopping trip. There was wild cannabis growing in the hedgerows and this vast monastery was built into the rock face. Monk graffiti was painted on the bare rocks, mantras and prayers in white, yellow, blue and red. Sonam's monk contacts again ensured some private escorted views of sacred inner temples and after leaving the site, we climbed a steady steep stepped pathway to a sacred spring. A large prayer wheel marked the site, and entertained children spinning the wheel whilst waiting for their parents to fill their large containers with this blessed water. Breaking all the rules of an English traveller abroad, I drank the spring water straight from the tap, confident that the abundance of

Tamshing Goemba

prayer flags, monks and centuries of blessings, kept this water as pure as could be. Indeed, it was delicious, and we returned many times to fill up our own drinking bottles and large containers for Sonam and Wangchuk to deliver to their families. I was excited to be meeting a friend of Julie, (my inspiration and who had visited Bhutan a couple of times over recent years) that evening. Pema and I had become Facebook friends following Julie's visits and I was absolutely delighted that we were able to meet up. Bumthang in Bhutan, is a district, rather than a small town, and so it was extremely lucky that I happened to be staying a mere few miles away from her home, and only a few hundred metres from her previous home. Facebook Messenger enabled us to communicate and she collected me from my hotel and together we ventured into the small town centre. In a tiny family run restaurant, we were the first guests of the night. In a private room, secured by a curtain, we caught up with who exactly we both are and why we had felt a desire to be friends. We were watched by The Bhutanese Royal family, photographs of whom were proudly framed around the room and on calendars. Pema volunteers for a charity 'Renew' and runs the Bumthang woman's textile and tailoring cooperative. As soon as we met, we knew we'd get along. I do love almost everyone I meet, but every now and again I meet special souls where we almost don't need to talk as a special connection or knowing exists between us. Pema was one of those people. Despite not having met before, we shared so many life experiences, good and bad along with both having a strong sense to improve the world and the lot of those around us. Pema told me that she wouldn't eat honey. Her husband however, who works in the Department of agriculture, does eat honey, and frequently bought local honey to give as gifts to guests and for when they travel. It caused great concern to her and we discussed why it may be a sin, and how a more natural way of keeping bees and extracting honey may not be considered so sinful. Like most people around the world, awareness of beekeeping methods is limited but I was still so baffled as to why the Buddha would consider eating honey such a sinful practice. Christians and Muslims treat honey as medicine and so therefore a precious gift from God. Buddhists believe that all of creation is equal, so the killing of living thing is considered a sin. The impressive rivers in Bhutan have never been fished, and no animal is slaughtered on Bhutanese land. Instinct is a powerful thing, and it turned out that we had an awful lot in common. Both mothers and step mothers, also on our second marriages. Pema is a very busy lady, working as a social worker, looking after her family with six children, also volunteering the charity 'Renew' helping victims of domestic abuse with a crisis centre, and an emergency refuge. Set up by Her Majesty, the Queen Mother Sangay Choden Wangchuk in 2004 there is also a craft workshop where women create crafts to sell to locals and tourists. I learned about the problems families experience here, which Pema appreciates are relatively mild compared to the suffering of some other women around the world. She is passionate to 'empower and enhance' women's lives, teaching them that an 'odd slap' or shove is not acceptable for merely serving lunch or dinner later than expected! Comparing the suffering of women around the world was an interesting discussion, both of us appreciating how lucky we are in our own circumstances. There is a drug and alcohol problem here in Bhutan, despite it being officially the happiest place on the planet. Introducing television, and bypassing cabled telephones straight to mobiles in the late 1990s has thrust the previously protected kingdom to be exposed to all that the rest of the world has to offer, good and bad. The young people are questioning the purpose of culture and heritage and using substances and television as a means of escape. Leaving Bhutan is not an easy option due to transport and educational limitations. Yet, visitors like myself were increasingly being drawn to this Kingdom, in search of our own glimpse of and celebration of happiness.

6. APIS LABORIOSA, ROCK BEES & KABIRAJ

THE FOLLOWING MORNING, WE MADE AN EARLY START

to visit one of the Cooperative Beekeepers, Kabiraj. We had to drive down a tricky track across a ditch into a field to his beautifully painted farmhouse. We'd picked up Navin to introduce us. Straight away Kabiraj was asking me how many hives I had and how long I'd been beekeeping. Navin's 150 hives over 18 years and Kabiraj was three times as many hives and 20 years experience! 'What on earth could you know with a mere 8 years of beekeeping and dozen hives?' he asked. Feeling slightly out of my depth, I repeated that I was here to learn and understand what is working in Bhutan and take back to the West lessons on how to live in harmony with bees and the landscape. The bees in the West are starving and sick and I wanted to find out how we could heal them. Relaxing, he spoke confidently and told me all about the bees they have, where he puts them, what the crops were and how the cooperative was restricted as they didn't produce enough honey to be able to export. Kabiraj is a very smart business man and beekeeper. His personal collection of honeys from both his own bees, Apis cerana and the stingless bees was impressive and he generously poured out large quantities for us to taste in bowls. Sonam politely joined in, I felt guilty for subjecting him to all this sin! Kabiraj told me of how the bees came into Bhutan from Kashmir, introduced by the Swiss businessman Fritz, who is the founder of Panda Beer in Bumthang. He spoke of how the bees in Bhutan don't suffer with any diseases. The main cause of death to colonies is predators, in the form of bears. For his bees to make the prized 'high altitude' honey, he needs to transport the hives to meadows 4000m above sea level. There they forage on white clover, wild cherries, roses, mint and rhododendron. The flavour was exquisite and quite unlike any honey I had ever tasted before. With only 1-2% of their bees being lost over winter each year or to predators, the initial 5 imported colonies has now expanded to over 1000 which are spread across central Bhutan. The Beekeeper's Cooperative has one wax foundation making machine used to make their own foundation for their hives. The machine had to be shared amongst the beekeepers which certainly wasn't ideal. It would be of concern if the wax had to be imported from other countries, increasing the risk of pathogens and chemical contaminants. The Bhutanese self-reliance and isolation from other countries' bees is helping to keep their colonies so healthy and strong. Kabiraj had encyclopaedic knowledge of his and the country's bees, telling me of all the locations of apiaries, bee farmers, the challenges and successes of the bees in Bhutan. His initial suspicion of why I was visiting soon disappeared as he realised that I was in his country to observe, learn and, where possible, help. My experiences from speaking with various bee experts and visiting bees in South Africa, Oman as well as Europe helped, as I could share their experiences and we both expanded our knowledge. The exportation of their surplus honey was an issue. The quantities were too small to warrant large scale export and the weight and transport issues made it difficult to support the domestic or tourist market. Meanwhile they had a growing number of healthy and happy bees producing an abundance of honey. With so much to discuss, he insisted that we stay for lunch and his wife produced the most delicious spread of rice, vegetables and the traditional dish of cheese with chillies. It caused great amusement to watch my reaction when eating this particular dish.

Kabiraj.

In the hotels, chilli is minimised to accommodate the more delicate palete of westerners. As a mother of sons where competitive chilli eating is one of our family hobbies, I impressed the group by managing to eat a reasonable quantity, for a tourist! From then on Wangchuk ensured that the hotels added chillies in all my meals. The trick I found was to be grateful to be offered first serving and to make the most of the abundance of cheese, avoid the seeds and mix with large quantities of rice! Kabiraj continued to share his knowledge, offering samples of the stingless bee honey. This was the same honey I had purchased at the airport in tiny glass bottles for 300ngu. Leaving several hours later with hands full of honey and stomachs full of rice and honey, we bade our farewell, watching their sons playing cricket in the field next to 30 beehives. Tamzhing Lhakhang is a beautiful temple round 5 km outside of Jakar on the Eastern side of the wide and rapid river. Sonam drove us as we chatted about bees and we teased him on the amount of suffering he was subjected to eating all that honey. Only a couple of other visitors were around and so we could enjoy the temple garden courtyard and inner rooms in peace. Wangchuk explained all the wall paintings and Buddha statues to me, ever more confusing with so many new names and stories to absorb. I loved this temple and we re-visited it before I left the area. It was established by Pema Lingpa in 1501 and said to be the most important Nyingma goemba (Nyingma literally means ancient and is the name given to the oldest tradition of Tibetan Buddhism. This school was founded on the first translations of Buddhist scriptures from Sanskrit into Old Tibetan in the eighth century) There was an iron cloak that we could wear and carry three times around the inner temple

Anyone who achieves this would be granted good health and blessings. We watched as the other visitors struggled past us. I refused the challenge on my first visit, still exhausted from my travelling, but on the next visit I was blessed with strength and energy to achieve the challenge.  Wangchuk and I then walked along the river to meet Sonam by the sacred spring water fountain to fill more water containers. Unknowingly we passed a colony of wild bees, attached to a concrete wall holding up a bank. A bee had flown out and stuck in my hair. I did fluster a little, it sounded so big and loud! Wangchuk managed to disentangle the bee and it returned to its cluster tucked away on the rock face beside the road. They were very aggressive, whenever we approached one or more flew back out. We decided to return with the car and my veil! These were the Apis Laboriosa or 'Rock bees'. Our colony was only 2 metres above the road and despite the cold temperature, still living in the North. Recent research identified another sub species of Apis dorsata, Apis laboriosa that was found at the higher altitudes. I wasn't equipped or prepared to take a sample bee for analysis this time around! Nicola's report had mentioned how Apis dorsata colonies were under threat as tourists were expecting, and

Rock bees.

paying good money, to view such harvests. In my planning I had requested seeing the bees but stressed how I wasn't interested in watching a harvest. Still recovering from the shock of my close encounter we came to the end of the track, to a classic rope bridge, slung across this terrifyingly fast river. Adorned with flags it felt like a majestic walk to cross this river. A photograph of it captured the wonder and awe of Bhutan for my family and friends at home. Prayer flags are a common sight across the Himalayas, and I was interested in the various shapes, sizes and positioning of the flags. Anywhere that could be thought to attract negative energy has flags hanging, the prayers and intent from those who wrote the passages on the flags, and the hanging, blesses the area. When someone passes away, a cluster of tall flag poles are erected in as high a place as the family can reach. The larger the number of flags, the greater the blessings. I loved seeing the 'Om mani pae mey hun' (the prayer for compassion) hung in many a rear car window! After crossing the bridge we found a small craft shop filled with hand woven scarves, rugs and the traditional 'gho' and 'kira' for men and women. There are many varieties of patterns, each from a local tradition and fabric familiar to the area they were made. Usually woven using cottons, wools and originally nettles. Sadly, due to the cost of hand weaving less expensive and manmade textile ghos are imported from India to supply the demand for the wearing of the national costume in most official roles. I bought a couple of the flag mantras for 60nu each and we bade our farewell. Collecting more water from the blessed spring, we saw the same young monk with his long queue of bottles to fill. Young children were spinning the prayer wheel setting off the bell.

APIS DORSATA 'The Giant Honey bee' is common throughout southern and lower altitude Bhutan. Similar in size to Apis laboriosa,(the Giant Mountain Honey bee) but with a brighter yellow abdomen. Apis laboriosa tends to be darker across the abdomen and found across the higher altitude areas of Bhutan. Apis Dorsata or 'Rock bees' are around the same size as our European hornet and I must admit to being quite shocked that I would encounter one so closely. Apis dorsata are the bees that we see on documentaries where tee shirt clad Nepalese or Indians climb ladders and cut away large pieces of comb. Traditionally, villages would have a 'honey hunter' who would take charge of this role. It would be at great risk as these bees do sting. Many days of preparation would be required to ensure a good harvest and most of the village would attend to lend a hand catching falling honey comb, making ladders and forming a human chain to pass any harvests along. Many honey hunters die, falling from great heights in their attempt to harvest the sweet delights of this wild bee. Driving around the Himalayas it is a common sight to see dark clusters of bees attached to their exposed comb high on the rock faces. These bees migrate, moving from higher to lower altitudes throughout the year in search of nectar. Their abandoned honey comb can easily be seen on the rock faces, or, hanging from high temple windows and roof tops. I encountered the Apis laboriosa in Bumthang along the roadside.

Lunch with Kabiral.

7. KUNZANGDRAK GOEMBA

Y SLEEP PATTERN STILL HADN'T SETTLED, WAKING OFTEN

hearing scratching noises above my head. Eventually sleeping and awaking naturally, I slowly prepared for breakfast at 8. Feeling tired and wondering if I should ask for the afternoon off, I needed some down time, alone, to think and write. The cloudy start cleared to another beautiful clear blue sky and I was eager to take the hike promised by Wangchuk. How could I resist the temptation of a hike to a temple? After an omelette and yogurt breakfast, we set off driving, this time turning right as we left the hotel. This is the main road to Mongar yet looking more like the lanes around my parent's village in Dorset. We drove out of Bumthang along my first experience of the mountain roads, narrow, bumpy, and very, very twisty as they hug the

mountainsides. Sonam is a calm and experienced driver and his former monk career obviously gives him the patience of a saint! There's no such thing as 'road rage' here, mind you there really are very few cars, or lorries. When a car does appear, it seems that everyone just carries on, breathing in a little and somehow the road is wide enough! The national speed limit is 40kmph but that's a far off goal as going much more than 20kmph would terrify any tourists! We twisted and turned on the tarmacked road, pine needles covering the edges and thankfully not passing, or meeting, many cars. We climbed the mountain sides and it was reminiscent of a Swiss valley, small farm houses scattered up the impossible slopes. Cattle and beehives in between wooden hay stalls and terraced rice paddy fields. I could understand how a Swiss businessman could feel this would be home. We passed the Burning lake entrance, marked with a large sign and prayer flags. The Burning lake is a very sacred spot where the founder of Bhutan, Pema Lingpa, discovered some of the hidden treasures in a cave below the water. Carrying on we passed a nunnery, perched high on a mountain, later dwarfed as we continued to climb. Turning off the main road and taking a dirt track, next to a drinking water fountain, we then started a mildly scary twisty climb up the mountain, 'Zig Zag' hill in Dorset now appearing very tame. This road reminding me of the Bolivian You-tube videos where a bus packed with passengers appears around a corner on a dirt lumpy

Burning Lake

track not wide enough for more than one vehicle. There were a couple of moments when I couldn't hold in a gasp, but Sonam is a careful driver, and I noted the white silk scarf tied around the steering wheel was giving us the extra protection needed to arrive at our destination. The road was treacherous, dusty clay and rock, single track and climbing steeply up the mountainside. We drove upwards for at least half an hour – put into perspective when Dorset's zig zag hill can be passed in no more than 15 minutes, when following a tractor!

The destination was Kunzangdrak Goemba.

The road ended at an equally scary car park, where one other car was already parked. Pine needles were coating the road and woodland floor, looking beautiful but treacherously slippery! With the sun out, and at around 3260m we began our 45 minute ascent to the temple, built by Pema Lingpa in 1488. We stopped to admire the view a few times and enjoyed the slight plateau as we approached the first chapel. A row of meditation cottages on our right for monks on their retreats, a three-year, three month and three-day retreat is part of their

training. After entering the Temple gates, a few of the last flowering marigolds were being visited by a honey bee, Apis laboriosa, found to thrive over 1000m. The first temple, Wangkhang Lhakhang, was up a couple of steep flights of stone steps, themselves balanced on a narrow ledge very high up above the valley below. There were a couple of monks to greet us and we walked around the first temple, spinning the prayer wheels as we went. Thankfully, it wasn't until after the visit that I learned that the buildings were all built by angelic beings, without using a single nail! The views down to the valley from the wooden decking,

Kunzangdrak Goemba.

Wangkhang Lhakhang.

suspended in mid-air, were spectacular. Removing shoes and hats, we entered the temple where a statue of the Buddha of compassion the central, surrounded by painted Buddhas, Chenresig, Guru Rinpoche and his disciple Namkhai Nyingpo. All were adorned with flowers, beads and donations of paper money. Butter lamps and sacred water, along with the hanging fabrics representing Buddha's earrings, body and dresses with masks filling the temple. Every inch of the walls was painted with lotus flowers, Buddhas and mantras. The three of us sat on small woven rugs and meditated for a blissful time. I imagined the Buddha of compassion and did my best to repeat the mantra, Ohm mani pai mei hun. Then I thought of Greg and my sons at home, asleep as they were six hours behind, my eldest away in Singapore, and my parents, sending them all love and imagining tight hugs. The time was so special, my hands were tingling, and I knew that this time would be committed to memory, ready to be re-experienced whenever I needed or wished. Only with the wind and the whispering chants of Sonam, we sat blissfully for longer than I thought we'd be able. Then in the distance Indian music could be heard, then voices. As the voices grew closer, we felt the vibrations as the group walked the 'Kora' path around the temple, spinning the prayer wheels as they passed. The rattling wheels vibrated the temple and I felt blessed to be sat high on a cliff ledge surrounded by sacred prayers being cast to the winds around me. The door opened and in came a Lama and 11 disciples. The monk showed them in, and with their beads and the Indian music playing on the mobile phone, they each blessed the statues and paid their respects. Wangchuk had by now opened the large flask of tea he had carried up the mountain and poured us each out a cup, complete with a tetley tea bag! Considering how close we were to Darjeeling and Assam, I did find this amusing. Apparently on pilgrimage from India, with the Lama, we were honoured to witness worshiping and the unveiling of one of the sacred treasures. A conch shell in a locked safe, covered in silk, but opened directly in front of me. The room was filled with the chanting of the twelve as Wangchuk, Sonam and I drank hot tea. After completing the chanting and each of us touching and blessing the relic, it was locked away again, and the monk told us to follow the group to the higher temple where a further relic, the stone anvil with Pema Lingpa's footprint embedded in it, and a group of wooden woodblocks used to print prayer flags were kept. We reached the temple before the pilgrims, and were glad to leave the rickety wooden structure, together. As we left descending the steps, we took a drink of the holy water that seeps from the cliff. I was most surprised that the monks had thought of all the needs of tourists with a loo positioned on the cliff side between the two main temples. It made Glastonbury festival's 'long drops' look positively shallow!

We returned to the lower temple, after using the 'mountain side facilities', flushed away using the sacred water constantly seeping from the temples. Enjoying another cup of tetleys in the china cups also carried up by Wangchuk, and a Kwix snack each that I'd brought with me, I explained that we could place the biodegradable wrappers in the pit marked 'Biodegradable' outside the entrance to the temple. The other pit labelled 'non-biodegradable' would apparently be emptied when full, by monks, the waste carried down the mountain then to the road and village, perhaps to be collected and disposed of somewhere that I have no idea where. Recent years I often find myself thinking, there's no such place as 'away'. Remote mountain communities like this certainly make one more aware of rubbish, the slightest waste has to go somewhere, but exactly where, in this pristine and perfect Kingdom? I can't imagine a corner with a deep hole filled with rubbish, marked from above by the usual seagulls. The walk down was easy, and pleasant with the Tang valley laid out before us. We passed the pilgrims resting part of the way down and we paused by some prayer flags when I shared Kate Bush, played on my iphone. Neither Wangchuk or Sonam had heard of her. We listed to part of 'Never be mine' and then I told them about the story of her live gigs. It was quite a special moment to be on top of a Himalayan mountain, with only birds, wind and Kate Bush! A couple of minutes was enough, when in such a peaceful place silence seems to be the most respectful sound. I was lucky that Greg and I attended her opening night in London in 2015. I was still wheelchair bound which was a real advantage in obtaining tickets! I had spent months gradually increasing my tolerance to sound and although wearing ear plugs for most of the concert, during her encore of 'Cloudbusting' I removed the protection and sang along as loud as I could. I'm sure Kate would have loved this spot, a grassy mound and the spreading valley below with the minute nunnery marking a mountain top. Turning the car around in the now quite full car park was worthy of a gasp or two, then we descended, thankfully not passing anyone driving up. We passed a farmhouse with beehives wrapped in old blankets and kiras. They must have been at 3500 metres, and definitely appreciating the extra warm layers.

I was pleased to know we'd be stopping at the Burning Lake. My friend Julie had been taken there by Pema for her birthday, butter lamps lit in her honour. We had this sacred sight to ourselves and were able to climb down to close to the waters edge. The river is very fast flowing and I could now appreciate the true wonder of Pema Lingpa's dive. These days any divers on purpose or accident, would surely be risking their lives. Fast currents and eddys were all around, with large cliff overhangs hiding deep trenches and caves. The bridge crossing the river was again adorned with flags and tiny stupas filled any shelf in the rock face.

tsa tsas.

STUPAS are structures used for meditation. They can be found at the highest points of mountain passes, or where roads or rivers meet. It is customary to walk around the Stupa clockwise many times in meditation. Larger ones will house relics. The mini ones we saw will often be made of clay containing the ash remains of loved ones. I loved the idea of my remains being sculpted into tiny stupas and placed in my favourite places around the world.

8. PICNICS

HEN TRAVELLING IN BHUTAN, TOURISTS PAY A TAX

which puts off many an unwary traveller. This 'tax' does in fact include all of you accommodation, and three meals a day. Your driver and guide are also included which made my 'tax' extremely good value. As a lone traveller, I also paid a surcharge to cover the additional costs of my trip. I wa extremely lucky that I was able to sit in the front passenger seat on ou journeys. Perhaps as an 'older woman' Wangchuk was respectful and felt h should sit in the back. I was extremely grateful as I do suffer with trave sickness and the twisty roads would challenge the hardiest of stomach: When visiting temples or people near to my hotel, we would return for lunch usually a hot meal of rice, vegetables, cheese and a selection of meats. I wa happy to eat vegetarian, but the perception is that Westerners like to eat mea When invited to stay for lunch with beekeepers, Wangchuk would call th hotel letting them know our plans. On longer days out, we had the joy c picnics. After visiting the Burning lake Sonam pulled over near som woodland across the road from the Nunnery. The ground was soft an covered in pine needles. Mats were laid out and huge flasks filled with rice an a selection of cooked vegetables and meats were spread out. The flasks a similar to the Indian tiffins used to feed workers. Throughout my stay we ha some delicious picnics in pretty impressive locations. Always with a view, ar always just the three of us. Bhutan is populated by just 700,000 peopl and visited by around 8000 tourists a year. This really is a place to escap from humanity. Although the landscape is splattered with prayer flags ar temples, the roads are empty, passing only a handful of cars over a few hour

The forests really moved me. The mountains are covered in trees, most have never been managed and so gigantic ancient trees with 'beard like' mosses hang from them. Each year on the King's birthday, school children and landowners plant 700,000 trees to maintain the 75% forested area.

As the sun moved across the sky, it quickly chilled in our forest clearing and so we began the drive back to Bumthang. Wangchuk became very quick at spotting bees and apiaries pointing out beekeepers beside the road in a large apiary. They were removing the syrup feed from the hives and closing each hive up ready for winter. The temperature was already cold out of the sunshine, yet bees were still flying in and out with pollen and nectar. We stopped to watch as the syrup-soaked straw was removed. By placing straw in the syrup bees had something to stand on whilst feeding and it minimised the risk of drowning. No fancy liquid feeders are available in this remote land. Old plastic containers were used above the brood boxes.

Picnic together.

I was concerned about the feeding of bees. In the West it is common practice to remove all of the honey and replace it with sugar fondant or syrup to give the bees enough food for the winter months. I strongly believe that a diet of sugar is detrimental to the health of bees. I then learned that sugar is not grown in Bhutan, and so beekeepers import it from India. The likelihood that it is organic sugar is very small.

9. URA

E HAD ANOTHER
DAY TO FILL
BEFORE A PLANNED

meeting with officials at the department of agriculture responsible for the beekeeping projects. We had been trying to alter my plans so that we could drive further East to Mongar and Lhuentse. I had heard of rock bees living on a cliff face outside the famous weaving village of Lhuentse, and was intrigued by the remoteness of villages and towns further east. I also have a love of textiles so knew that I would enjoy meeting weavers and seeing the richly embroidered silk favoured by the Queens and Princesses of Bhutan. The road to Mongar is still mainly unpaved, extremely twisty, making the journey which looked achievable on the map, a much longer adventure. Accommodation in these areas is also scarce and we'd need at least two days to travel there and back. Wangchuk came up with an alternative itinerary travelling as far as we could on day excursions from Bumthang. I was awoken again in the night by some odd scratching noise above my head. It could have been a bird, rat or squirrel. I decided that it wasn't a spider so went back to sleep. At breakfast a group of 6 French speaking Spanish, French and Swiss were also sharing the dining room. I chatted a little with the Spanish gentleman. They had been taken to a dance festival in a remote village nearby- it had been on all weekend. I managed to have enough wifi strength to do a short 'Live' Good Morning post, keeping all my friends and family up to date with my exploits. I was sharing the bandwidth with these additional visitors so I got used to intermittent access. We left just after 9.30 am headed through Bumthang town centre to join the main road to Thimpu. twisted through the forest, up the mountains, very narrow, bumpy and twist

I had seen the road from the aeroplane when I came into land at Bumthang. At the high point a Stupa and many flags marked the border between Bumthang and Trongsa districts. After around 1 hour, we turned left for the new highway to Ura, referred to as the 'Bypass' a slightly wider tarmacked road. This was also twisty but markedly better than the previous road. We travelled for almost another hour then pulled over above the valley of Ura. Wangchuk and I left the car and walked the easy track down into the very quiet village. We passed new born calves, the crematorium and the closed temple. I mentioned that I now needed the loo. Not the best place to find oneself short! We knocked on a few doors but no one was around. A family of stray dogs family followed us with their fluffy pups, then we came across ponies and more cows. Thankfully we then passed a farm house where two ladies were outside weaving and spinning.

An ideal opportunity, Wangchuk asked if we could watch, then asked if they had a loo I could use. I was taken inside the farmhouse, up a steep set of stone stairs, then inside two flights of wooden 'ladders' to the top floor. Unlocking a room which looked large and bare, mattresses folded in a corner and a log stove in the centre. She turned on some light switches then led me to a balcony outside with a stainless steel sink and another locked door which was a small loo – western style – with a flush! I left a fairly generous tip, and returned down to the garden. Not to miss an opportunity, rugs had been laid out and displayed, thankfully one was to my taste and budget, what I had been looking for my bathroom at home. It was made by 'Tshering' one of the sisters who was sat spinning green sheep's wool. She had made the patterned rug and was happy for me to photograph her with my rug. I handed over the 'bargain' 2000NU. They were both stressing that it was far cheaper than the shops would sell it for! Then they pulled out belts hoping to up sell. 🐝 We were then offered to come inside for butter tea- 'Ara'. It would have been rude to refuse, so in we went. A large stainless steel pot was filled with water, tea and butter then whisked up using a hand blender before she poured it into two china mugs. A pot of 'crispy corn' and 'puffed rice' both hand made were offered so I had handfuls of both, to help soak up the tea. As soon as I'd drunk some tea, Tshering topped it up. I didn't learn and continued to sip faster than Wangchuk who managed to get away with only one refill! 🐝 Then she pulled out the strong stuff, perhaps to entice me to buy a belt, or another rug. It was distilled by herself using wheat, also home grown. The pinkish colour was from infusing dried sandalwood herbs into the brew which was left for 2 months. It was warming as it passed down the throat, not too strong, and quite pleasantly flavoured. By now I needed the bathroom again, and as I'd left my bags downstairs with the almost unfinished 4th cup of tea, I was grateful of the gift of a bunch of loo paper I'd left after my last visit, thinking of any other passing tourist. 🐝 61 year old Tsering and her sister run the house as part of the 'Home stay' program in Bhutan and had had a busy year with three western guests through the summer. Apparently, I was much more inquisitive and open minded than her previous guests!

We spent a little more time outside watching and learning about the weaving process, taking photos before walking back through the village into the valley then up the other side to meet Sonam and our picnic lunch. A beautiful grassy plateau was set up for lunch and we shared rice, chilli vegetables, potatoes, beans and mushrooms, all in the metal tiffin pots I'd so liked previously, and all sat on my new rug. As the boys packed away the emptied pots, plates and rugs, they insisted I rest, so I lay out in the sunshine on my rug and listened to the birds and the wind. No aeroplanes, or rumbling traffic sounds. Rather than take the original longer road back to Bumthang we returned along the highway, passing the new stupa being built. I was fascinated by the wooden bamboo scaffolding, reminding me of

reenactments of the Pyramids being built. There was a road side café/ hotel and I enjoyed their facilities remembering my granny's advice to never pass a loo! This was a hotel for locals stopping off as they cross the country. We passed families employed to paint white lines on the new roads, babies on backs, sons holding long ropes whilst the father painted the lines. Parts of the road were unfinished and were merely a sandy track in between large piles of rubble.

Tshering pictured here with 'The stronger stuff'.

10. DEPARTMENT OF AGRICULTURE

WAS ABLE TO HAVE A LATER

BREAKFAST AND LATER

start as my appointment with Dr Sonam was at Wangchuk 11am. We left the hotel just after 10am despite the department of agriculture buildings only being a couple of miles away. Sonam was not as calm as usual today, and I even witnessed impatience as we were held up by a very smokey fumed lorry, who only after much hooting of horns, let us pass! We arrived at the department buildings, like a palace, and above the airport, just around 10.20 so passed the time watching the bees around the apiary in front of the buildings. The bees were flying as the temperature was warm, and sunny. We spotted two helicopters fly over the Dzong, heading North, maybe they'd turn around to land at the airport? It caused much interest as Bhutan only owns two helicopters, used by Royalty, military and emergency rescue. We spotted hygienic behaviour by the bees, rigorously cleaning other bees before letting them enter the hives. We also saw drones, flying out and entering the hives. Then Wangchuk spotted a red bottomed 'bee'- on closer inspection it was a fly, but mimicking a bumble bee of some kind. Just before 11, Dr Wangchuk came outside to greet us and my guides left to collect more Holy water from the spring. I was shown upstairs into the meeting room. The layout of the interior stairs and galleries was very much like a temple, doorways covered with a fabric curtain. Entering the meeting room, a large 'U' shape of tables went around the room and I sat at the far end. Sonam went to fetch his assistant Prasad,who is managing the beekeeping project across Bhutan. Their prime objective is to produce organic honey. As they realise they simply cannot compete with the bigger honey producers of the world they need to find their special niche, which is undoubtedly their

— 37 —

country's clean air and nutritious soil, protected from chemical and pesticide use. 🐝 When Apis mellifera were introduced into Bhutan they were aware that it may cause problems with the native Apis cerana bees, and so there is are strict guidelines about the areas in which Apis mellifera apiaries can be housed- the north of the country, and central parts, and along the highway. In these areas there are Apis dorsata and Apis laboriosa and so no risk of the cerana coming into contact, or conflict with the Apis mellifera. 🐝 They have had a problem with the education of Bhutanese that it is not a sin to take honey and they are thinking that in Bhudda's time the bees were always killed to extract the honey. In the south traditional cerana beekeeping is still maintained, but in a way that the bees are no longer killed. 🐝 Tregano bees are also managed in such a way that the bees are not sacrificed, by using a syringe to extract the honey from the cups. I shared that the Indonesians have been adding coconut shells to the hives as a 'super' to collect honey protecting the original colony. 🐝 I also asked about hornets and they say in the south it can be a problem but the locals use cow dung around log hives which I thought may be interesting with regard to hornet deterring after I explained Heather Mattilla's research in Cambodia. 🐝 To develop organic honey they are aware that the practice of feeding sugar will have to be stopped. Currently they have pockets of organic honey. High altitude honey is also of interest, however hives have to be carried by hand and so it is not currently viable on a commercial scale. 🐝 I asked about a tree I saw in Ura with the red blossom and full of bees, I was told it was actually pomegranate 'Punicaceae' flowering March to July in Thimphu, Punakha and Trashigang, yet we saw it in Ura in November! Flowers yield a red dye. Originally cultivated for flowers and fruit and now naturalised in dry inner valleys. 🐝 Apis mellifera introduction is limited to Haa & Paro, and was introduced into Chukka but it wasn't successful, I am not sure if that was due to farmers or location. 🐝 I learned that Bumthang has greater forage for bees than many other areas. 🐝 Traditional beekeeping in the south is from 800-1800m above sea level, and the beekeepers there have no need to feed the bees sugar as there is forage all year round. I was given contact details of the beekeeping official in that area so that we could arrange a visit.

Much of the funding for the beekeeping project was from the Swiss Company 'Helvetas' and founder of Panda Beer. It wasn't limitless and so they were still needing support for research and were very keen to work with Universities around the world to improve their facilities and knowledge about the various species of bees in Bhutan. So far no research had been carried out on the solitary or Bumblebee population of Bhutan. I instantly thought of Dave Goulson from Sussex University and his extensive studies on bees, and in particular I was concerned about the impact of introducing a new species of honey bee had on the native pollinators. I also learned that there are Apis florea bees in the south of Bhutan, the same species I had seen in Oman. As Sonam and Prasad opened up, there was so much to see and learn in this country, I was wishing I had more time to follow up all of the beekeeping contacts they gave me, and hoped that I could encourage further research to take place.  Coincidently, Pema's husband worked in the neighbouring office and is an agricultural nutritionist. Dr Gyeltshen was happy to meet with me and Pema introduced us whilst their son sat patiently, watching closely as we

Pema with her son.

all spoke in English. His nutritionist knowledge was astounding, especially when I have been used to western doctors for both people and animals focussing mainly on treatments and not nutrition. He was surprised to learn of the practice of feeding bees sugar, knowing as he did how harmful it would be to their long term health. Not wanting to miss an opportunity with a nutritional expert, I asked him about the wheat here as I am not able to tolerate eating it at home. He told me all about the wheat grown locally and slow cooked which releases particular polysaccharides that make it easier to digest the grains. In the West we use modified seeds that are then cooked too fast, making it difficult to digest. Many processed foods here are from India though so it was best that I didn't eat wheat whilst here. The hotels mainly used locally grown foods, and I found there was plenty of buckwheat used instead of wheat and I was absolutely fine eating that.

At the Department of Agriculture with Prasad and Dr. Sonam Wangchuk.

'Dr. Gyeltshen's nutritionist knowledge was astounding especially when I have been used to western doctors for both people and animals focussing mainly on treatments and not nutrition'

11. DAMCHOE & TRADITIONAL MEDICINE

UE TO THE SMALL POPULATION, ALMOST EVERYONE

knows each other, either through family, school or work. It takes the 'six steps away' idea to a whole new level. Realising that I was studying herbal medicine, Wangchuk contacted a school friend of his, who was now a Doctor in the local hospital. After a short telephone chat, it was decided that I could meet Damchoe in the hospital the following day. The hospital was in a converted palace, donated by a former queen of Bhutan and so a beautiful, as well as historic building. We entered, spinning the gigantic prayer wheels, familiar with any public building in this Buddhist Kingdom. I was firstly struck by how quiet the hospital was. A handful of people around reception areas and long empty corridors, this was in outpatients! Every hospital in Bhutan has a conventional 'Western' medicine department, AND a traditional medicine section. Both are fully equipped with clinics, dispensaries and beds for those needing them. I was shown Damchoe's office, a large room lined with a display of dried herbs and medicines. A couch was against the other wall with a consulting desk and a few chairs. It looked perfectly normal and extraordinary at the same time. Damchoe had excellent English and so was able to talk me through the services they can provide. Every patient can choose how they want to be treated, either with traditional or Western medicine. Since the introduction of paper money in 1963, and then TV and internet in 1999, Bhutan has rapidly been exposed to the wonders of commerce and the rest of the world. Sex and the City, Kardashians and various talent shows from USA to Indian Bollywood are all well-known and familiar. Despite the availability of global TV, the rapid introduction omitted a gradual development through the 1930s black and white films or any of our so called 'classics'. Star Wars, Doris Day, Hitchcock or Clark Gable were all completely unknown. The same was for music, anything twentieth century completely missed by these people who still wear traditional dress in public. It seems that medicine is the same. Watching American hospital dramas created a demand for a similar style of medicine. The very young demand to be global citizens and not be left in the dark ages.  The very old, also excited by a new money-based society, wanted paracetamol and migraine medications as seen so frequently advertised on the satellite TV channels. There did remain however a strong tradition of using tried and tested medicines, made from their home-grown herbs and in a tradition dating back farther than anyone can remember. Towards the end of my travels around Bhutan, I had a few days in the capital Thimpu, and visited the National Traditional Medicine hospital with its adjoining University. One of the largest modern buildings in the city, and so the country. It was bustling with young students, and the hospital clinics had full waiting rooms. I loved the modernity of it all. The signs offering a dream menu of available traditional treatments. The health service in Bhutan is free to all, certainly a great place to be ill! Across the country I met with many people, rural farmers, working women, children and families. One common theme was the spiral of additional medicines required to combat the effects of Western medicine. Blinded by TV advertising, the pressure to embrace the new rather than stand up and share the old traditional ways was overwhelming and I sincerely hope that this pattern decreases. TV has also brought a demand for foreign imported food. A country who until so recently grew their own food or were fed by parents who still did. The new young King and Queen, both Oxbridge educated, are eager to maintain Bhutan's quality and uniqueness as a place not of the past, but the country that can cope with the future. Each family has been given four acres of land to enable them to grow their own food. I was surprised to see the variety of crops grown in this diverse country. Rice was common everywhere, along with chillies, covering roofs during my stay, drying out in the Autumn sun. Buckwheat was another staple, great for bees, and gluten free diets. Acres of apples were grown across central Bhutan, exported to India, then citrus, pineapples and bananas in the South with cardamom and all manner of other herbs and spices. Knowing my love of teas, Wangchuk took me to a small farm shop in Bumthang where I could buy a large bag of tea leaves. They were dried and for 500gm it cost me around 30p. He called it 'shimja' tea, leaves collected from the mountains by goat and cattle herders. It wasn't until I returned home that I opened the bag and tried it out. A few leaves produce a very dark, maroon colour tea. Astonishingly it is not in the least bit bitter, despite its strength. It looks fruity with a slight aroma of blackcurrant leaves. It is slightly drying on the palette, and maybe like an Earl Grey without the bergamot! I tried researching on the internet for the tea, I was becoming addicted to my cup of Bhutan every morning and eager to know why, and what it may be. Nothing was coming up

Damchoe's clinic.

so I asked Wangchuk for more information about this Himalayan tea treasure. The dried leaves were very dark and crisp, some quite large and twiggy. Then I remembered Damchoe, my very own Bhutanese medical herbalist Doctor. He replied that it was 'hypericum hookerianum', on further research a member of the St Johns wort family. I told him of how we use St John's wort, which he found interesting as they didn't use it for depression or sleep disorders at all! When I researched this herb, it has proven to have powerful antibacterial properties. Yet Damchoe said In fact they don't even use this herb for medicinal purposes. He further researched my possible tea leaves and came up with Shingja growing in the lower altitudes, and Osyris lanceolate growing in the higher altitudes. Wangchuk told me this tea was what he would drink high in the mountains with his herds of cattle, so I knew this was a high

Hospital corridor

altitude crop. Connecting all the people and tales I heard during my travels, I remembered hearing of someone's mother, who I met at her mountain top farm. After I returned home she plunged into one of her depressive episodes. Her sons had been buying her western medicine for her migraines and depressive periods. Depression is one of the main health problems afflicting this land of happiness. It is most common with older uneducated women. With their new wealth, and her simple life, it was a sign of pride that her sons could buy these new drugs for her. During December and January she was hospitalised with severe depression. I couldn't help but wonder if this simple tea, now rejected for PG tips and Typhoo, would have helped keep such women as her happy in their simple lives in this beautiful land.

'The new young King and Queen, both Oxbridge educated, are eager to maintain Bhutan's quality and uniqueness as not a place not of the past, but the country that can cope with the future'

12. TRONGSA

UE TO THE LONGER DRIVE, WE SET OFF EARLIER,

8am for Trongsa. The plan was to stop at Wangchuks parent's house on the way for tea, then visit the tower museum and the Dzong, before returning to their house for lunch. I was getting used to the twisty roads and even beginning to recognise certain features, such as the electricity power station before the steep incline out of Bumthang and the point where we turned left for Ura but continued right this time to Trongsa. Much of the road was still dust and mud and there were occasional 'depots' where the contracted Indians were mixing tarmac, using diggers to move rubble and stones, and other machinery to scrape away the sides of the rocks to widen the road. We drove through some beautiful villages that looked so interesting with traditional farm houses, schools and little shops selling the fresh bettle nut that Sonam needs to keep awake during the drive. I made a mental note to definitely return to this section of Bhutan with more time. Forested valleys and extremely picturesque villages with monasteries and

Trongsa road

temples crowning the peaks around. The current Queen, a commoner, was from one of these villages, and the Chhume valley was absolutely stunning. We drove past a school where all the children were sat on stools outside, as if all the classrooms had been moved into the fresh air. Fantastic inspirational quotes, in both English and the native tongue of Dzongkha were painted on the walls. I regret not stopping to take photos on the way as it was dark on our return. We stopped at the highest point Yotong La, where a Stupa marked the centre of the dusty road. Anticipating increased tourism, a block of public toilets was being built, but not yet finished or useable. Although the road was a high point, it had been cut into the mountain, and so cliffs rose either side of us. Trees covered the southern side and I couldn't resist scrambling up a worn pathway to take a look at the view. Frost still covered the trees and undergrowth, with cobwebs glistening in the morning sunlight. I caught a glimpse through the mists of the Black Mountains, said to be rich in minerals but mining of them strictly forbidden, protecting the area for future generations. Topped with snow and circled in more mystic tales of good and evil, I love how the tales of angels building temples and evil forces battling over this sacred land are included in everyday conversation. No one bats an eyelid when talking such things. The line between what we would think of reality and mysticism is a very fine one here, like in Oman, yet both are extremely religious countries. I find it interesting when comparing this to Western tales of dragons and angels which are fixed firmly into the realms of fantasy. As we descended the other side of the pass, a dramatic, yet slightly disheartening view of miles of twisty roads lay before us, slowly descending over 1000 metres to Trongsa, which at this point was still hidden. I knew we had a long drive ahead of us. We eventually pulled over, able to have a fantastic view of the ancient central capital of Bhutan. We still had quite a descent but across the valley could see the freshly cleared road to Punakha across the valley. Wangchuk's family farm was an impressive building (Sonam called it a castle!) built above the road and with far reaching views down on Trongsa, and up into the mountains behind it. The house is a traditional Bhutanese home, of two stories, where the family live on the upper floor. We stopped for tea, then headed off down the valley to visit the Tower of Trongsa

Monks in Trongsa.

Royal Heritage Museum. Sonam dropped us off on a sharp bend, so we could walk through woodland (Wangchuks former walk to school). Arriving at the tower, we were still high above the valley and the Dzong, which are often romantically shrouded in mists. I really enjoyed this museum filled with tales of battles, gurus, monks and kings. The highlight for us all was to see the Raven Crown, worn by kings for official ceremonies. Unlike most museums with this quality of exhibits, it was only the three of us, with small groups of tourists following behind as we walked through the rooms of Armour, cloaks and statues. All our phones, cameras and bags had to be left behind in lockers, such are these treasures. The gardens around the tower were filled with large colourful blooms, we were much lower than Bumthang and so the temperature warmer, almost in the twenties. Gigantic butterflies flitted between the flowers, and in the heat, we walked further down the mountain to the impressive Dzong, argued to be one of the most spectacular sited in Bhutan. The town was far more touristy than I was expecting, many small shops along the roadside, similar to a Cornish fishing town in Summer. Pot plants decorated the buildings and trade was being done all around. On entering the Dzong, we were lucky to have heard that bees were living on the roof. We were directed to a covered pole with bees coming and going high above the courtyards. Young monks were carrying tea urns up the white steps, their bright red gowns looking fit for a painting. The main festival season was approaching, we were so lucky to then walk into rehearsals of the dances and singing to take place in a few weeks time. The five day 'Trongsa tsechu' would take place just after I left in December, so I was very excited to be able to witness these special dances without all the crowds! The tsechu would culminate with the unveiling of a

— 47 —

giant *thangka* (a painted or embroidered religious picture.) Thangkas offer the viewer great blessings and healing and so Bhutanese will travel great distances as a pilgrimage to view such an unveiling.

Wangchuks' family picnic.

By the time we had explored the Dzong, we were ready for lunch. Grateful of Sonam driving us back up the hill to Wangchuks family home, a fine picnic was laid out waiting for us in their garden. The hospitality I received was heartfelt and heart-warming. His family were shy to speak in English, yet they all understood me. I was given a tour of their grounds, all planted with enough food for the whole family, no need for supermarket shopping here. The most astonishing part of the house I was shown was the family temple. Just as we would have a large room with chairs directed towards a television, the largest room of this house, (and as I learned most homes) was a dedicated temple.

Wangchuks family temple.

This was the only temple I was allowed to take a photograph of. The sun streamed in through the windows on to the wooden floor. The family Buddha statue draped in offerings, traditional drums covered in tasselled drapes. It took my breath away. On the floor was a rolled-up mat, an Uncle, living as a monk, stayed here. I could have stayed with this family until thrown out, the location was idyllic, and I felt for Wangchuk leaving this tranquil home for a life in the city, taking tourists like me around his country.

13. LAST DAY IN BUMTHANG

HE DRIVE BACK FROM TRONGSA WAS LONG AND TIRING

for us all, lightened when after asking, 'where can I see Yaks?', I pointed out a large 'fluffy cow'. Sonam slammed his brakes on and quickly reversed and to my delight we found a small herd of Yaks. I didn't feel a trip here would be complete without seeing them. The Village Lodge in Bumthang, with the friendly staff, was feeling like home and awaking on my last day was marred with sadness. The boys would be heading off to make the long drive back to Paro, ready to meet me off the plane the next day. They needed to leave by lunch time at the latest, we all felt sad at even a short separation. The day was sunny and bright and I was happy for them to leave me in the town shopping and I would walk back to the hotel, eager to take photographs of the chillies drying, and some of the interesting characters we drove past every day at the small shops, bars and farms. Before setting off, another trip to the sacred spring was needed to fill containers to give to their families on their drive home. Wangchuks family in Trongsa, then a planned stop in Thimphu for the night where Sonam's family lived. They would each sleep in their own beds for a night before setting off early the next morning to Paro to meet me off the plane. I remembered the Yak herder ladies outside the temple near the spring, so asked to be dropped there whilst they filled their containers. I spread my custom amongst the ladies, buying one gift from them each and was particularly excited to have found a perfect gift for Julie. As well as a Buddhist traveller, Julie is an accomplished artist, her paintings of her travels to Bhutan being part of my draw to the country. I found a tiny bronze paint pot shaped as a flower with each petal a covered receptacle for paint. The covers were held in place by a pin which when lifted, opened the petals and their lids. It sat in the palm of my hand, but in Bronze, weighed a fair few pounds! Once alone, I wandered the streets of Bumthang with its market, and large variety of shops. I love being alone and wandering, without a care for time. I looked for and found Pema's women's craft shop, tucked away along a back street. The scarves were wonderful colours and made perfect gifts. I then found an excellent crafts shop filled with perfect gifts for everyone, 'Lhakey Handicraft'. I knew that my weight limit was going to be stretched, so had to think carefully about singing bowls and the large phallic statues which were to be seen everywhere. So much to admire and make one smile. One of the joys of being so far away from home, is that we forget about how 'connected' we are. All was well until I tried to use my visa card to pay for my chosen items. Like any western shop, the familiar card facilities and signs were on view but my purchase sent a shock wave across the continent to some office in the UK, certain that it can't have been me wanting to shop in Bhutan. There is excellent phone signal in Bhutan, but no agreement with my phone service provider. I had purchased a sim card here to use with a mobile wifi gadget and was given a simple phone and sim card for use in Bhutan by my travel agent in India. Needless to say, neither I or Barclaycard had thought through an effective way of checking my identity whilst in Bhutan, and despite the shopkeeper kindly topping up my sim card so that I could call back, yet another phone operative after I would be cut off, to start the process again, and again. Barclaycard were certain that my card was clear and available to use, I could even use the cashpoint machine positioned just outside on the street. Yet I couldn't. Eventually the shop machine did accept my card purchase, but the cashpoint machine failure concerned me. I had brought what I thought was plenty of cash, yet if I was to adequately tip the hotel staff, and my guides, as they all so deserved, I was going to have to be far less generous without access to a cashpoint. I began to think about the deals my own government were trying to negotiate as we were leaving the European Union. This fiasco took time, during which the sun sank behind the mountains. I realised how protected and cared for I'd been by my guides, always making sure I was out of the bitter sub-zero temperatures, reserved for the shady afternoons.

My long walk back to the hotel, in the shade, without my warm coat as it was so warm earlier in the sunshine, ended up being a rapid march, with my scarf wrapped across my face to allow me to breathe, it really was that cold. The hotel restaurant with its large stove covered in pebbles for us to warm our hands on, seemed quiet without my guides. A new group of tourists had arrived and filled the room, chatting in their own language and drinking wine. I suddenly felt very alone and homesick as well as that awful feeling of being broke.

'I pointed out a large 'fluffy cow'.
Sonam slammed his brakes on and quickly
reversed and to my delight we found a
small herd of Yaks. I didn't feel a trip here
would be complete without seeing them'

Spring water.

14. TURNING FIFTY

T WAS VERY COMFORTING TO CHAT TO GREG ON

my birthday eve, and I slept well after spending a very long time wearing all my heavy items and packing everything else. I left as generous a tip as I dared, apologising and saying how they were all worth so much more than I could give them. I really didn't like this feeling of lack, after so long feeling abundant and generous. My flight was after breakfast, at a reasonable hour as it was the plane returning from dropping off fresh tourists from Paro. The large group offered up one of their guides to escort me to the airport and alone but optimistic I awaited in the field with all the strangers waiting for our plane to land and for us to board. There is a small departures lounge with security check in. My bags, and I were scanned, and my passport was checked, then I joined all the other waiting passengers out in the field. The plane landed and I smiled as the tractor carried their bags from the plane to be collected. I felt like I was a local here in Bumthang, everything being familiar, and being alone like a *real* traveller. Once the plane was emptied, we all filed back in to go through security again to access the footpath to the plane, the other side of a wire fence. A Lama and his assistant were also boarding. I was delighted when he sat across the aisle but in front of me. Such a smiley happy fellow, blessing the captain and the crew, and chatting with the Slovenian passengers told to sit in front of me to balance the weight. Greg had given me a beautiful silk scarf, with bees woven into the fabric, before I left, I'd worn it every day so far on my trip, but today it was really mine, and I was turning fifty. The flight was spectacular. Clear skies gave the most wonderful views of snow-capped mountains, forested hills and valleys with the road not only cutting through Bhutan, but also the sides of the mountains. I learned later that Wangchuks family all waved as my plane flew over, I looked down on their farm, hoping I'd soon be back. My mood was the opposite of the previous night. I'd had my down, now I was up, quite literally on top of the world and looking forward to seeing my dear friends, my Bhutanese boys, and the adventures we had yet to come. Once we landed in Paro, I was the last to leave, before the Lama. I

On top of the world.

wanted to thank him for being on my flight and blessing it for us, as my previous flight was quite scary. I also had to tell him I was fifty today and so feeling extra special to spend time with a Lama. He blessed me and presented me with a beaded woven bracelet. I left the plane grinning ear to ear, and I've worn my bracelet every day since. Already happy, I was caught by surprise when I was met outside by Wangchuk and Sonam, presenting me with white scarves. I hugged them both, feeling so very blessed indeed. I had no plans for the day, trusting that my expert guides had it all arranged. They dropped me off at my first home in Bhutan, the Village Lodge in Paro, this time overlooking the other mountains, but an equally delightful room complete with a balcony. I was able to freshen up and was reunited with my other suitcase that they'd kindly kept for me whilst I was in Bumthang. I opened some of the Birthday cards I'd brought with me and gifts, feeling really special and loved. 🐝 Our first stop was a temple Kyichu Lhakhang one of Bhutans oldest and most beautiful temples. An inner courtyard was planted with flowers

Chanting

and houses a mural of King Gesar of Ling, the popular Tibetan warrior king whose epic poem is said to be the world's longest. We could hear chanting of monks to our right, and to our left women singing and drumming. I was so excited and asked Sonam if we could enter and see the women. Initially he thought not, but I was hopeful, and after a short disappearance, Sonam returned saying that we could go and watch. We walked bare foot through male temples with monks chanting and praying then found ourselves looking through a doorway into a room filled with women, two monks and a yogi, all sat cross legged on the floor towards the richly adorned statue. They soon stopped for a tea break, then as one of the ladies stood up to share the refreshments, they invited me in to join them. I sat on the floor amongst them as they giggled at my appearance and look of wonder and gratitude. The Yogi commented on my age, saying that I looked 'only 30 and so I must be a good soul and that after this life I would go to a good place'. What better compliment could I wish for? I sat with them for over an hour whilst my guides sat patiently in the doorway, and various groups of tourists came and went looking in on us. The ladies were chanting from ancient texts, I tried to assist my neighbour by turning the page when I thought it was needed. They drummed on choeddrums, sang, chanted and blew bone trumpets. I later learned that these were treasured human bones from a particular tribe of tall people in Southern Bhutan. Longer legs were prized, and it was a great honour to have ones thigh bones re utilised as a red painted trumpet. Closing my eyes I felt that I was in a thousand places and a thousand times, all at once. Stonehenge, Hawaii, Africa, New Zealand, South America, anywhere my imagination cared to take me, or was it memories of where my soul had been? My eyes filled with tears; I knew that this was exactly the perfect place for me to be on this auspicious day. I wondered if my instinct to travel to certain places was to return to past life locations, perhaps to heal both my soul and the place. Whatever the reason, I knew that my instinct to travel to Bhutan for this birthday was absolutely spot on.

That evening, my hotel was hosting a traditional dance for guests from another hotel. Feeling both privileged and naughty, I sat above the stage on a balcony with guides and enjoyed the performance. During the day, both Sonam and Wangchuk kept disappearing on odd errands, as one or other of them took me to temples and restaurants. Unusually, Wangchuk was insisting that I would need a cup of tea and we should meet in the lounge area overlooking Tigers Nest. I was absolutely flabbergasted when their scheming resulted in a huge birthday cake being brought into the dining room. We all had a slice and shared the cake with the staff and other guests of the hotel. We then visited the hotel's temple, where a butter lamp was lit in my honour. I don't think I have had a day so filled with smiles and happiness in a very long time.

15. BEES AND TEMPLES, PARO

E PACKED A LOT IN ON MY BIRTHDAY INCLUDING A VISIT TO

the Dzong, where, as if to celebrate my fiftieth, there were bees. Prior to my visit to Bhutan, I had heard of bees nesting on the outside of Temples, dramatic photographs of ornately painted windows with vast comb hanging below, covered in bees. This interested me on many levels, not only the species of bee, Apis dorsata, but also the location. Why would bees be building their combs on these buildings?

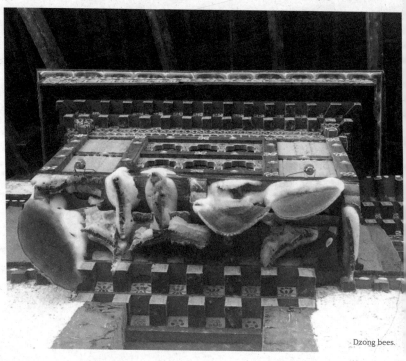

Dzong bees.

I soon learned during my trip, that most temples have bees. Before building a Dzong or temple, monks and Rinpoches spend much time meditating on the best location. Paro Dzong's name is 'Rinchen Pung Dzong' meaning, 'fortress on a pile of jewels'. It was built on the foundations of a monastery built by Guru Rinpoche, completely by hand and without a single nail. In 659 the Tibetian King Songtsen Gampo decided to build 108 temples in one day to pin down a giant ogress who had caused the dowry of his Chinese Princess wife, to become stuck in mud. The temples would pin the ogress down forever and also convert the Tibetan people to Buddhism. Her shoulders and hips are pinned down across the four main districts of Tibet, and the temples in Bhutan pin down her troublesome left leg. Kyichu Lhakhang (my birthday temple) pins down her left foot, and her left knee is pinned down by Jampey Lhakhang in Bumthang. Throughout my trip I learned of meditation caves where gurus had visualised the building of certain temples across the country. The Bhutanese also raise prayer flags wherever they feel there may be 'bad energy'. Rivers meeting, a sharp bend in a road, high peaks and any river crossing all now benefit from an array of prayer flags. Like Western churches, the citing of a religious building often reflects ancient beliefs in the sacredness of that particular location. Important sites are also known to sit over geopathic stress lines, around the globe, which I know are attractive to bees. Although Buddhist monks may attract bees as they do not harm them, refusing even to use abandoned wax comb for candles, bees could really be drawn to these locations because of the earth's energetic vibration, geopathic stress lines frequency matches that of bees at around 250Hz. The Paro Dzong bees had left, migrating south for the winter months. They had left their huge wax comb hanging from the window ledges. As it dried out and crumbled to the ground, the monks simply swept it up. Like vegans, Buddhists don't eat honey as they feel it's cruel to take the bees food, which I completely understand. With conventional beekeeping and exploitation of bees, removing all their honey and replacing it with sugar, it is indeed detrimental to the bees. If there is a gentler way of working with the bees, taking small amounts of honey, without killing or harming the bees, then surely this could be acceptable to both monks and vegans?

Not only are Dzongs and Temples beautiful buildings, but the monks in their red robes make for stunning images. What I particularly loved were the printed signs around the buildings, written in both Dzongkha and English. 'Please don't ring or fiddle the bell', which makes a bell rope all the more tempting to fiddle with. My fascination with toilets was particularly tickled by 'ORDER other than Dratshang, no body is allowed to use Dratshang's toilet'. (Dratshangs are the monks) All the signs are beautifully painted. Rubbish bins in the grounds have 'Fill me up' painted on them. The bee combs were hanging way above my head so not able to inspect closely, we raised suspicious looks from the Dzong guards as we walked around the outside of the Dzong, straining our necks as we looked upwards.

As we returned to the hotel, we continued along the road towards Tsento, past the luxurious five star Amankora resort hidden in dense woodland, and on to the derelict, but being restored, Drukgyel Dzong. The afternoon sunlight streaming onto it making one of the most picturesque images I'd seen. Perched on a small mound, it protected Bhutan from Northern invaders, reminding me of Dorset's Corfe Castle. I'd missed the bees in Paro, but we would be hot on their trail as we prepared for our journey south.

Drukgyel Dzong.

Paro monks.

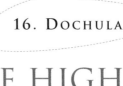

16. DOCHULA

HE HIGHEST POINT BETWEEN
THIMPHU AND PUNAKHA

is a pass known as Dochula. This place and name wouldn't have made any impact on me if I hadn't read Emma Slade's book 'Set free' on my journey to Bhutan. Emma was a British Banker who ended up rejecting her high living lifestyle and designer shoes to become a nun based between Bhutan and Canterbury. Her chance meeting with a Lama in the temple at Dochula transformed her life, and although initially my itinerary didn't include this site, I was so excited that Wangchuk's updated 'Bee plan' included it. At 3140m it gives spectacular views north to the snow-capped Bhutan Himalaya. We were blessed with a bright blue clear November sky and our drive there along the modern road via the capital Thimphu was picturesque. The Queen funded 108 chortens to be built here in 2005 as atonement for the loss of life caused by the battle against Assamese militants in Southern Bhutan. With a son in the military, I felt a close connection in this magical place. There is a temple above the chortens, and the café which was the main focus point for the sudden appearance of tourists. Having spent most of my time alone with my guides, it was extraordinary to suddenly find myself with people from all around the world, including a couple from Wales who I shared a cup of tea with. talking with them made me so grateful to be alone. In their late sixties, the married couple hadn't agreed on where to visit in Bhutan. She was very much like me and loved the treks and temples. He was less active and not such a cheerful soul. I could see that her trip was a massive compromise for the sake of keeping her husband happy. Sometimes we need to be selfish, for the good of our own health and wellbeing. I plan to return here, with my husband, but then I will be happy to be led by what he wants to see and do as my own needs will have already been met. Since Emma Slade's visit, a charge is now made to enter the temple, but I felt it had to be seen. It's recently been completed with modern painted murals. Quite extraordinary after seeing so many ancient ones, this painted history of Bhutan in such fresh bright colours and with so much gold was quite different. I was hoping to find Emma's Lama, from her book I had imagined a tiny simple temple perched on top of the hill, with an elderly Lama sat in the dark, with a single sunbeam highlighting his red gowns. Instead this was a vast modern temple, with statues and paintings and many thousands of Buddhas. A young smiling monk blessed us as we entered. It turned out that he could well have been Emma's Lama, another lesson in not being attached to expectations, or maybe even having the courage to sometimes ask questions!

Dochula view.

17. PUNAKHA

HE DRIVE DOWN FROM DOCHULA TO PUNAKHA

was spectacular, and terrifying! I found it hard to believe that this was not only a major route, but *the* major route across Bhutan. Tourists and Bhutanese who chose not to fly travel in small buses and taxis making this part of the country pretty busy for such a narrow twisty road. The Royal Botanic Park is positioned a short distance below Dochula amongst a vast forest of rhododendron, alder, cypress, hemlock and fir. These magnificent forests would look incredible during spring when the rhododendron is flowering. I made a note to plan a return visit then. There is a white chorten built in the road to ease the negative energy from a number of road accidents in this area.

Punakha.

Chortens always have to be navigated clockwise, making this almost a large roundabout placed on a narrow mountain road. The extremely beautiful views combined with twisty narrow roads really is a recipe for accidents. As we descended, much of the road was still in shade, and despite the warmth of the sun, the ice was still solid on the shaded parts of the road. Waterfalls are often seen pouring down the mountainsides and particularly near the inner bends. Signs forbidding car washings were frequently seen in these areas, but it was too icy to contemplate stopping to take a photograph, let alone wash the car. It was only an hour driving down the mountain to Punakha valley, but it felt long enough, I was relieved to arrive in what seemed like another country completely. The temperature was distinctly warmer, in the twenties which I really wasn't expecting. Tropical plants surrounded the valley and children were playing in the sand along the river's edge. It felt like a Summer holiday destination. The Dzong here is possibly one of the most photographed and was built in between two very fast wide rivers, Mo Chhu and Pho Chhu. It was the second Dzong to be built in Bhutan and was the capital until the mid 1950s. All of Bhutan's kings have been crowned here and it is the Winter residence of the monks. Guru Rinpoche had foretold the building of this Dzong, describing its location as by the hill that 'looks like a sleeping elephant'. Wangchuk talked of the importance of building a Dzong at the 'confluence' of the two rivers, a new word for me! To access the Dzong, we needed to cross a beautiful covered wooden bridge and I was delighted to see evidence of bees above the main entrance to the Dzong. Again the bees had left, but there were still plenty to be seen around the beautiful flowering grounds. This Dzong had many tourists visiting and preparations were being made as a rumour that the Queen was planning to visit gave an air of expectation. We chatted with a couple of monks here, hoping to glean more information about bees and their relationship with monks. They were just bemused as to why we would want to use beeswax for anything.

Punakha bridge.

18. THE DIVINE MADMAN

JUST OUTSIDE PUNAKHA, BACK TOWARDS THIMPHU,

there's a village called Sopsokha with the temple Chimi Lhakhang a short walk beyond.

The village is surrounded by small farmsteads where rice and buckwheat were being harvested by hand. We walked through the fields and rice paddies passing farmers laden with huge bundles of buckwheat. As we passed the houses, I noticed an extraordinary amount of painted decorations, all of them variations of phalluses! All sizes, shapes and colours were on full display, then the gift shops appeared. Tables full of hand carved wooden phalluses painted in all colours, with added keyrings, bottle openers or with added wings and painted faces. We had arrived in the land of the Divine Madman. Lama Drukpa Kunley (1455-1529) is a Tibetan born saint who had an extraordinary way of spreading Buddhism. As a yogi he would travel around

The phallus symbol.

Tibet and Bhutan behaving outrageously, singing bawdy songs and humour, with the purpose of attracting the common man to his message. There are numerous tales of his exploits including seducing the wives of his hosts, hanging blessed threads around his penis to give him 'luck with the ladies', and urinating on Thangkas. It is slightly uncomfortable walking past so many explicit paintings and sculptures and yet it can't help but raise a smile. He used his behaviour to 'symbolise the discomfort that society expresses when facing the truth'. The phallus symbol now is painted on buildings across Bhutan to protect fertility and respect Drukpa Kunley's wisdom. I saw many hanging on the corners of houses with wings on to add extra protection. He really is 'the people's lama'. Not surprisingly, this village is very much on the tour guides route and so again I mingled with all

nationalities as we walked from the village to the temple. An avenue of trees was quite beautiful with a large Bodhi tree at the entrance to the temple grounds. Buddha is said to have received enlightenment while meditating under such a tree and so many of the temples have one in the central courtyard. This one was particularly beautiful and with the afternoon sun streaming through its branches it was a perfect place to sit and meditate, or people watch. Many visitors come to

Bodhi tree.

this temple built in 1499 to receive blessings. Kunley's cousin built it to commemorate the Lama's subduing of the demon at Dochula. Murals inside depicted the Lama's life and a queue of childless couples waited patiently to receive blessings from the Lama's wooden and bone phalluses. Pregnant mothers can collect a baby's name from a bowl. I steered well clear of all the relics and monks in this chapel! We had a long day ahead of us and so we were all happy to retire to our respective rooms and get a good night's sleep.

19. TSIRANG: APIS CERANA

E NEEDED TO SET OFF BY
8.30AM TO GIVE US TIME

to find the beekeepers and return before dark. I slept reasonably well, awaking refreshed seconds before my 7am alarm. I showered, washed my hair then did a live video as I watched the sun rise over the mountain from my balcony. Breakfast was scrambled egg, rice porridge, banana and water melon. It kept me going until lunch, aside from the rise balls purchased from a market stall around Norboogang. We headed south along a surprisingly busy and wide road. Passing the Wangdue Dzong, that was burned to the ground a few years ago and now in the process of being rebuilt. The sun was shining and the weather forecast was for temperatures in excess of 26 degrees C!

Tsirang District

Not long after leaving Wangdue and the main junction to Trongsa, we drove through the major hydroelectric dam works. For over 45 minutes we were driving through massive construction works with villages, schools and towns building up along the roadside to cater for the immigrant Indian workers and their families. The scale of the project was astounding; I knew my father would love it! The main dam building on the opposite side of the fast rushing river had tunnels and caves with men in fluorescent jackets looking like ants in comparison. After passing the dam, we twisted around the cliffs looking down on the rapids, huge boulders and frothy white water. Spiralling down we finally crossed the river on an iron bridge, welcoming us to Tsirang District. We began another climb, goats on the road side, instead of monkeys and cows, and I noticed white lines along the centre of the road. We soon came across a row of market stalls with ladies selling a vast array of pickles. and stopped so I could

Bridge in Dagana

look, hoping to find some honey. There was a selection on each stall, one lady told me it was 2000Nu per jar. Later looking at a photo, I see it says 300- but I don't know if that was the price. I bought some puffed rice 'balls', mildly sweet, which proved handy to offer to ladies we met later. Behind the stalls, crossing the fast flowing river, was a long bridge, swinging. Wangchuk and I started to cross it, then spotted an elderly figure carrying a basket on her back coming across. We waited until she was close then asked if I could photograph her with me. She agreed. I'm finding that no one smiles when you photograph them, yet they smile constantly before and afterwards! Turns out she lives alone in the 'shack' on the other side of the river, aged 85 and has to come across to collect water as she has none on her side. Her husband is in hospital in Thimpu, and her sisters' children are keeping an eye on him for her.

After snacking on our rice snacks, and taking photos of a young girl on a stall, we continued our drive, the road now zig zagging upwards. A junction directed us left and up towards Tsirang, whereas straight on, along the river and later crossing it at Sunshine bridge towards Dagana. A destination for a future visit. After around half an hour, at 10.15am, we arrived at another junction, outside Tsirang, the entrance to a school to our left and a lady sat on a wall waiting for a bus. She was local but wanting to visit Thimpu. I gave her a rice ball and took her photo- she had lovely gold jewellery on. We turned left onto a narrow tarmacked road which twisted and turned FOREVER! Not knowing who, what or when we were going to arrive I just had to enjoy the views, which were amazing! Deep drops into the valley with terraced slopes for rice, citrus, kiwi and all manner of crops. The area was lush, fertile and splattered with small holdings. There were fewer flags or stuppas which was partly due to the fact that we were further south and so more Hindus and mixed religions. This was the area that the King had been giving land to families and so Bhutanese from all over the country were settling here. Finally, after around 45 minutes, the road ended, and a large fern covered archway announced the Government agricultural official's buildings. Our contact from Dr Wangchuk met us and directed Sonam where to park. The road continued steeply up the hill, but without tarmac. As we then set off on foot, a truck passed

Road to bee house

Tsirang 'honey house'.

us filled with sacks of smoked cardamom. -Delicious!! We walked along the track, then turned off through the narrow paths in between the terraced farmland.

Through cardamon, citrus, bananas and a stunning selection of woodland, we passed by remote farmhouses until we came across the familiar sound of buzzing and I knew we'd found the bees! Two houses had balconies with log hives, blocked with cowdung, mostly empty, then we came across the full hives, bees buzzing over our heads to the terraces. Then a hole in the outer wall of the house was pointed out – a real actual 'honey house'! They make a hive inside the walls of the house whilst they are building, purely for the bees!

Bee hive entrance.

— 70 —

Bee hive in wall of house.

This place really felt like heaven, the light, the scents, the colours, bees and the people. The honey house owner greeted us whilst his two small children giggled and stared at this strange group of people who had suddenly appeared to look at their bees. Using Wangchuk as a translator, I asked about what he grew, how many hives he had and how he takes honey. He had log hives as well as the wall hives

Log hive.

The log hives were hung around the outside of the house, just as northern Bhutanese had flying phalluses. Growing citrus and cardamom, amongst the rice and buckwheat, he was able to harvest from the bees twice a year; May/June and again in October/ November. All the hives were natural so no need for expensive extraction equipment. The comb was simply cut out, back to the brood layers, and then pressed. You could say it was the 'Bees' knees' honey as there was no filtering and so not just knees and wings were stored in his old 'Rock Bees' whisky bottles. The thick dark syrup poured easily out of his bottle as he let us taste this precious harvest. Each colony yields around 2kg of honey at each harvest and they always ensure leaving at least two frames of brood in each colony after harvesting for the colony to survive. Many of the honey houses here would have 6-7 hives over the summer, our chap Bal, had twelve. He agreed to sell me some of his honey and I was honoured to have been given a whole Rock Bee bottle for a very reasonable fee. We were then invited to travel across the mountains for another hour to visit more honey houses and farmers, but with over a four-hour drive back, we simply didn't have the time. This was somewhere I definitely had to return.

APIS CERANA The asian or eastern honey bee is slightly smaller, and lives in smaller colonies than Apis mellifera and has been used for pollination across Bhutan for many hundreds of years. They are easy to keep as far more docile and resistant to many diseases. They nest in cavities, and are often kept in boxes, logs or the traditional Bhutanese 'honey houses' where space in the walls is left during building.

20. DOCHULA MEDITATION & APPLE ORCHARDS

XHILARATED BUT TIRED

AFTER OUR MAGICAL

journey south to Tsirang, we were sad to be leaving Punakha on our way back to Thimphu. We felt that my trip had climaxed. We'd found the honey houses and what appeared to be perfect living, in harmony with nature, far away from the stresses of even Bhutanese life. I wasn't entirely looking forward to the drive back up the mountains but thankfully, as is often the way, the return journey feels quicker. As we approached Dochula, I was not only relieved knowing that there would be toilets, but excited as Wangchuk had planned a mini meditation retreat for me. I had completely missed the beautifully painted meditation caves south of the memorial chortens, during our last visit. Like all the other visitors, I was looking north towards the Himalayan peaks. Crossing the road, and moving away from the tourists is another restaurant, quieter and protected in the forest. (No queues for the loos here!) This leads you into the woods and a climb up to the meditation caves. I couldn't believe my eyes. This place was incredible, small hobbit style caves, open to the Himalayan mountain view with a stone seat and 'table'. The rear wall of each of the 11 caves was painted with a different god or goddess. They were built to commemorate the 60th birth anniversary of His Majesty the Fourth Druk Gyelpo (Dragon King). Lama Nima Tshering of Dochula Lhakhang says 'The caves will benefit those who seek for inner peace after a long week, months or years of tussle with worldly matters'. 'I can see visitors leaving happily after meditating for a short period. Though their happiness may be temporary, it gives me immense satisfaction that they are at least happy for a moment, in this place. Thinking so, my respect for Her Majesty the Queen Mother Ashi Dorji Wangmo Wangchuk grows even deeper," Lama Nima added

A beautifully painted meditation cave south of the memorial chortens

in an article for Bhutan today in 2017 when the caves were completed. The construction was made possible by the Queen Mother and as well as the caves, there are other structures all crafted into the shape of 60 to commemorate the birthday of the former king.  I was so excited, and eager to find the perfect cave, I settled on the god for energy, Chana Dojre also known as Vajrapani. I felt that energy is what I lacked during my poorly years, and now I was blessed with bucket loads of it. This was the God to thank. I felt like the

The god for energy painting on the cave wall.

happy Buddha sat alone in my cave, sometimes closing my eyes, and other times not resisting the temptation to just gaze at the most incredible view. I could have stayed here for at least the three days recommended to begin meaningful meditation, but even though I didn't have that long, Wangchuk and Sonam were considerate to leave me far longer than most of their charge would be happily left. It was hard to leave this area, so, like children, we clambered up and down the steps aiming to find all 11 of the caves and capture a photograph of each of their painted walls. No one else was there, either walking or meditating. What a find! Recharged and energised, I was ready to head back towards civilisation and tourists around Thimphu. Descending the mountain pass from Dochula there is a check point where everyone has to have their documents stamped. It ensures that officials know who has passed this point to the centre and East of Bhutan, and more importantly, who has returned safely. Anticipating the possible business from awaiting car loads, a market has sprung up with stalls selling refreshing juices and water and lots and lots of apples. As my visit coincided with a recent harvest, bags of apple were all around and the drive back towards Thimphu was surrounded in appl orchards. This region is famed for their fruit growing and export their excess to neighbouring countries. I was particularly intrigued as my husband Greg manages orchards and with my interest in bees I am always pressuring him to work in a chemical free manner. In the West it is common belief that it is impossible to grow apples commercially without the use of pesticides, insecticides and fungicides. It was this use of chemicals, sprayed early in the mornings to 'protect' honey bees, that prompted the first investigations on the effects of these toxins on our native bee populations. Bumbles and solitary bees make earlier starts than honeybees and so were often caught by the sprays. The Red mason bee is the most effective pollinator of apples and a single bee can do the work of 250 honey bees. 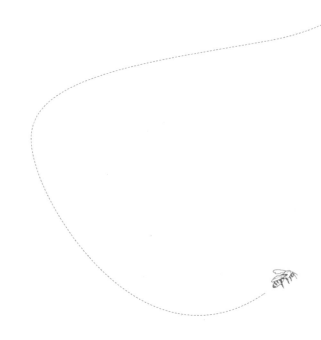 With orchards covering all the slopes, I gazed out of the car window, looking forward to returning with Greg and walking amongst these healthy pure trees.

21. THIMPHU

ANGCHUK HAD BEEN MAKING A FEW PHONE

calls during our long drive and arriving in Thimphu I was so grateful of the result. He impressed me greatly by securing a magnificent room with a view in central Thimphu. All the hotels I had stayed in were extremely clean and comfortable, but Punakha was less friendly, and for reasons I cannot now recall, I was disappointed and missing the spoils of The Village Lodges in Bumthang and Paro. I was tired and ready for a rest. There had been so much to take in during this trip and I was ready for some time to rest and record what I had experienced. 🐝 We drove through the city and parked outside the Kisa Hotel with its beautifully grand painted walls. The entrance foyer was more like a Western five star hotel and I hated admitting that I missed life's little luxuries. As I was shown my room, I was delighted to be high up with two walls of windows overlooking the city park. It got me thinking about our expectations. Driving across Bhutan over the previous weeks, I realised just how difficult it was to transport anything into or out of this mountainous region. Even the main modern road from Paro to Thimphu was narrow and twisty, limiting transportation to smaller solid lorries. Modern life has got us used to keeping everything we own new and fresh, re decorating and replacing furniture on a whim or a change of fashion. Here, furniture is made solid and long lasting from the trees cut in close proximity, and why would anyone consider importing replacement beds and wardrobes unless absolutely necessary? 🐝 I was ashamed of my own, and 'my peoples' disregard for sustainability. We think nothing of replacing furniture, taking our unwanted pieces to the charity shop to ease any guilt. Thankfully there is a demand for second hand furniture not just for those limited on budget

So much more ends up broken and in landfill as our expectations for 'perfection' increase. Closer inspection of my hotel room would leave a typical western tourist describing rooms as 'tired' or 'in need of upgrading' yet the room was spotlessly clean with a large comfortable bed, chairs and a dressing table. In my own home I have always been reluctant to re decorate, feeling once I've completed a room then it's done, hoping, forever. We have been so spoilt with the ease of replacement, and of disposal with no regard for the consequences. Thimphu is more like any city around the world. Cleaner than most and with a population of under 115,000, it still appears bustling and overflowing into an otherwise very rural country. Despite shops, banks, hotels and the other trappings of a capital, the nation's Buddhism was the outstanding feature. A large National Stupa attracted many visitors and locals

People sat by prayer wheels.

throughout each day. We joined them circling around, faster than some, slower than others. Pebbles piled around the edges helped the pilgrims to measure their rounds. Elderly sat by the prayer wheels gently spinning them. To my eyes I feared they were homeless, hoping for generous donations from this tourist magnet. Wangchuk explained that the elderly choose to sit here, making good use of their time praying for those without so much spare time. I was particularly excited to be visiting the huge golden Buddha statue positioned high on a mountain side, overlooking the city. I'd seen it from the plane flying back from Bumthang. Built to commemorate the 60th anniversary of the fourth King Jigme Singye Wangchuk in 2010, it also fulfils two ancient prophecies of the 8th and 12th centuries which stated that a giant Buddha statue would be built in the region to bless the world with peace and happiness.

The 51-metre bronze statue sits on a three-storey base and contains 125,000 smaller buddha statues. There is a park we could walk to with yet more splendid views across the valley and over Thimphu. Sonam and Wangchuk live in Thimphu and had fun pointing out their homes to me. Now in the nation's capital, I was certain that my need for more cash would be easily satisfied, with the Bank of Bhutan (BoB) along with cashpoint machines on every street. As we trudged the streets from one machine to the next, I enjoyed watching the central road crossing with a traffic warden in a little decorated building in the centre. His beautifully choreographed arm and white-gloved hand movements controlling the traffic from all four directions.

The 51-metre bronze buddha

None of the machines would accept my variety of debit and credit cards and so we decided to visit the main bank. Wangchuk was excellent at staying calm and translating my questions about bank transfers, paypal and other methods of turning a number on a screen to useable paper notes was asked. The extremely helpful 'foreign currencies' manager in his office upstairs, with his personal card machine and direct line to my bank had to explain that British banks don't have an agreement with Bhutanese Banks and so my cards could not be used here, other than in a shop displaying the international symbols of wealth we take for granted. Finally, Wangchuk knew how he could help, and we were driven by Sonam to a friend's gift shop, with the reassuring colourful stickers of all major global cards filling her windows and doors. Gifts purchased and making use of her 'cash back' facility, I was now feeling much more relaxed and excited that we could visit a grocer so I could buy the ingredients to bake my famed flapjacks. There were no large supermarkets surrounding or even in this city. Even though I chose not to use supermarkets at home, buying our vegetables from a local organic grower, and meat from butchers, I was still surprised that a city of this size, with so many visitors, was feeding itself from tiny corner shops and markets. My ingredients list was oats, sugar, golden or rice syrup and butter. The latter wasn't a problem, the hotel would have plenty. The other ingredients, despite me thinking they would be staples, were very difficult to find. Oats, not usually grown here were in small imported packages, far more expensive than at home in the UK. Familiar bags of sugar were nowhere to be seen, neither were the tins of golden syrup, I had presumed would be found in all corners of the globe! Improvising, I decided on coconut sugar, honey and the imported oats. Excited about using the kitchen in my 'home' in Paro and sharing my home baking, as well as having some cash in my purse, I slept soundly on my last night in Thimphu.

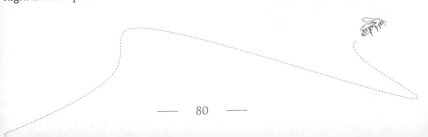

22. EMMA PEMA

AFTER READING EMMA SLADE'S BOOK 'SET FREE' ON MY

way to Bhutan, I felt moved to contact the author and let her know how much I enjoyed and was moved by her tale. Also, that I too was now in Bhutan and could relate to much of her story. It came as a great surprise to receive a reply to my email whilst I was still in Bumthang, giving me the number of her assistant in Bhutan so that we could arrange to meet. Our revised itinerary, giving me a night in Thimphu meant that we could indeed

Emma Pema.

meet up, so I invited her to join me for dinner in my hotel. There was much excitement amongst the staff that the famous English Buddhist nun Pema Deki, would be visiting the hotel. During Emma's transformation from a banker to a nun, she had been renamed by her jovial Lama, who giggled when he picked 'Pema', rhyming so perfectly with 'Emma'. I too loved the 'Emma Pema' name which helped to identify her both in the UK and from the many Pemas in Bhutan. We instantly had so much in common and to chat about. I wanted

to learn more about her charity, 'Opening your heart to Bhutan'. Through her own life journey and with her autistic son, she had a special affinity with disability and felt that she could help her beloved Bhutan by helping children here with disabilities. In the remote areas, facilities for physically or mentally disabled children is sparse and misunderstood. Deeply religious, as well as with strong faith in the mystical, afflicted children can often be labelled as 'cursed' or living as a result of trauma or bad deeds in former lives. Often children are kept hidden away, for fear of spreading the affliction. Then there is also the problem of ancient stepped buildings, bumpy roads and complete lack of available health care or treatments. Through simple acts of compassion, a basic in the training of any Buddhist, Emma's charity provides access to safe medical care, disability aids such as wheelchairs and prosthetics and basic amenities to isolated and rural areas of Bhutan. Believing that education is a means of empowerment, children are taught traditional crafts and skills to enable them to become contributing parts of their community. Her charity has no paid employees and is run purely by volunteers and donations from organisations both in Bhutan and across the Globe. During the build-up to my birthday, social media had been sending me reminders and suggestions to set up a fundraiser to celebrate. This charity seemed to be the perfect one to support, I was in Bhutan because I no longer needed a wheelchair, and often thought that even if I had visited during those disabled years I would not have been able to access many of the ancient and modern sites that I now could. I began to imagine how different my life would have been if I'd been born with Ehlers Danlos in Bhutan. So grateful of the funds my generous friends and family had raised to celebrate my birthday, I now had a connection with Bhutan, it's disabled children, and the wonderful Emma Pema. On the return drive to Paro from Thimphu, I could feel this magnificent trip drawing to a close. I didn't want this short drive to end, wanting to take in every sound, smell and sight we passed. The rapidly rushing river running alongside the road and the tempting road signs leading to mysterious places I wouldn't be visiting this time. One place we could stop at was the Iron Bridge, Tamchhog Lhakhang, built in the 15th Century by Thangtong Gyalpo. As a Tibetan saint and engineer he was thought to be the first to use iron chains in the construction of suspension bridges. He'd visited Bhutan in search of iron ore, and a trail of eight heavy iron

chain bridges were built across Bhutan from Paro to the far Eastern province of Trashigang. Unfortunately, the only surviving original bridge was washed away in the far east of the country in 2004. A replacement bridge was suspended across this fast flowing river for pilgrims and tourists to cross and admire the beautifully decorated chorten and small temple.

Iron Bridge, Tamchhog Lhakhang.

23. BUMDRAK TREK

ONCE I ARRIVED IN BHUTAN, I BEGAN LOOKING

through my guide book and at the detailed map my local bookshop had sourced for me. Taktshang Goemba, or The Tiger's nest, was already on my itinerary but after seeing it from my hotel window on that first night, I was intrigued by the lights I'd seen above and around it. I hadn't planned on a trek, although I did want to walk whilst in the Himalayas, after all, isn't that what one should do here? My guide book had a whole chapter on treks, many were several days long and required much advanced planning, but then I spotted the Bumdrak Trek, a short two day trek which included a trek down to Taktshang Goemba. My travel agent in the UK, Maria, acted on my excited message and put plans in place along with Wangchuk. It was decided that my last two days would be the ideal, and only, time to undertake the trek, and I had sufficient kit (walking boots and warm layers) with me so it was all decided. We set off from the hotel at 8am to arrive at Sangchen Choekor Shedra Buddhist college, ready for a 9am start. The road twisting up from Paro was the scariest yet. Wangchuk said that last time he was on this road it was on a bus! Even he was looking away from the edge, and I had to grab the hand rail a couple of times. Sonam just laughed! There were a few gasps and I was very relieved that there was no other traffic. I have since learned that taxis charge extra for this steep windy journey. The sun was coming up and so there were a few blinding moments as the road was invisible due to the glare. We stopped near the top at a look out point where we had a fine view of the airport runway. We could see a plane preparing for take-off, most likely to Bumthang. We watched as it taxied towards us, turning south then with all the engines at full blast, it tore down the runway, leaving the ground at around half way along. From our height the runway looked extremely short and it was a relief to see how little of it the plane needed. As the plane rose, it negotiated the nearby mountains and we could see it rise up and into the sky. A plane flying in the skies of Bhutan is not a common occurrence, with Paro as the only international airport and not open every day. During our trek I noticed how the sound of the wind blowing the prayer flags through the trees was the only background noise, no rumble of motorways or roads. There was building work going on, but even that was quiet compared to the noise levels we're used to in the UK. Arriving at the college, the monks were up and happy to open the temple for us to make our prayers for a safe trek. I also had the obligatory photograph standing next to their giant stuffed bear in the entrance hall. I was expecting a large group of tourists to be gathering in the carpark, but there was only us. At 9am Wangchuk and I bade farewell to Sonam, and only then did I realise that we were the only ones on the trek today! Luckily the route was clearly signposted as we climbed out of the college grounds into the woodland. We passed the monk's baseball and football pitches then a fork in the path guided us either to the meditation huts or the Bumdrack trail. Silence was requested as we passed the huts. There are many meditation huts and retreats high up in the mountains of Bhutan, specifically for the silence. A few birds and a butterfly joined us, but otherwise it was very quiet.

Meditation Huts

A Bumdrak monk.

There had been a forest fire the previous year and so much of the woodland was bare and blackened. It did mean that in between the charcoal tree trunks, we had a clear view of the valley below. I had naively thought that as we were driving high up to a temple to start our trek that most of our walking would be level, along the ridge line. Sangchen Choekor is at 2900m, lunch was another 500m climb and the campsite 1000m above us and so we had a challenging walk ahead of us. I soon felt the altitude compressing my lungs. My usual speed and brisk pace reducing to a panting crawl. I had packed the minimum I thought I could manage with but not knowing just how cold it would be at the mountain top campsite I had included a selection of woollies and thermals, along with my stainless steel 'drink-safe' filter water bottle and the remaining flapjacks. I'd also been given the tip to take a portable phone charger, which had already been useful, but it does weigh a fair amount! Now I wished that I hadn't brought anything!

The view into Paro.

'As we climbed into thicker forest, Wangchuk began whistling. Bears and tigers prefer to avoid human contact wherever possible, stories of being stuffed obviously concern them, and the whistling keeps beasts away from us, also preventing them being caught by surprise!'

Wangchuk had managed to travel light- a smaller ruck sack but heavy with water. He noticed me struggling and very kindly insisted on carrying my bag for me. It was still cold, and, in the woodland, we were still in the shade, so our coats and scarves were still needed. As we climbed into thicker forest, Wangchuk began whistling. Bears and tigers prefer to avoid human contact wherever possible, stories of being stuffed obviously concern them, and the whistling keeps beasts away from us, also preventing them being caught by surprise! This reminded me of when I had made a stag jump as I'd been wandering through woodland. The pathway was clearly marked by regular deposits from the horses that are the main source of transporting supplies to the temples and camp site higher up. Gradually the woodland became greener, and warmer as the sun was higher and the exertion of the climb warmed us up.

Looking up to Choethotse

It was a great relief to arrive at Choethotse Lhakhang built on a small ledge at 3650m. A table was positioned amongst tall prayer flags and the view was splendid over the Paro valley. Above us I could see another cluster of prayer flags, thankful that they hadn't been our destination. Little did I realise that after lunch, they were indeed on our path to the campsite! Our lunch of hot rice and vegetables had been brought down to us from one the resident caretakers of the campsite. He set off with us, but soon we were left behind as my coastal lungs struggled to maintain his hearty pace. The forests we walked through were filled with oak trees and rhododendron and I imagined would look spectacular in the Spring. Having watched the tiger documentary before leaving for Bhutan, I now took a keen interest in the droppings we passed, partly excited about encountering a tiger, but also knowing that we certainly shouldn't wish for it. Wangchuk told me of a school friend of his, a farmer, who went missing and was later found having been killed by a tiger. His wife was left to care for three young children, the fear associated with death through such a terrible incident left her alone as the thought of her and her children being able to transmit their bad fortune. She later committed suicide, leaving the children to be brought up by the government as yet more

— 89 —

bad luck was surrounding these unfortunate orphans. I thought of Emma Pema's work through education, and how perhaps her charity could have helped these unfortunate victims of tragedy. We stopped a couple of times to admire the view, eat flapjacks and for me to recover my breath. It was a great relief to finally reach the circle of prayer flags with the green tents only slightly beyond. It was still early afternoon and so we could relax on reclining chairs in the warm sunshine. Appearing to be the only guests, milky tea was brought to us and I enjoyed watching the passing locals! One lady was bent over with a huge slice of a tree attached to her back. She stopped for a

flapjack, and I should add now that most who tried them found them too sweet, an indication of how spoiled our palates have become with sugar. Later a man passed carrying a chainsaw. We learned that the lady's brother was meditating, and she had brought up the log for firewood. Then a Yak herder passed by, face aged by Himalayan sun and wind. He also tried a flapjack then went on his way. His yak herd were grazing below us. None of these passers-by had walking boots or ruck sacks, seemingly

Flapjacks on Bumdrak

walking in flip flops or soft shoes along a familiar route. It wasn't long before I'd recovered enough to go exploring. The cliff hugging hermitage of Bumdrak Lhakhang, built nail less by angelic beings, was accessed by a collection of vertical ladders at 3900m. The location is said to have been where 100,000 dakinis (female sacred spirits) used to frequent. Bum means 100,000. This really was a very magical place. The resident monk had an enviable seat overlooking the campsite, and the complete Paro valley, his simple living quarters were his home whilst he oversaw the meditation huts and blew the sunset and sunrise horn. We also climbed behind the camp to take a look at that view and find the perfect spot to hang my prayer flags from Bumthang. Wangchuk bravely climbed a tree overhanging the valley and we felt a suitable place for our prayers was secured. Watching the sunset, a small group of trekkers arrived at the campsite with their guides.

Bumdrak camp arrival.

Dinner was served by lamplight in one of the tents and my tent was filled with a solid wooden bed piled high with layers of blankets, yak skins and sheets. The covers were reassuringly heavy as I dosed off to sleep hoping that I wouldn't need to get up in the night! The temperature plummeted, I was awoken by barking dogs and a very loud purring sound between my bed and my personal loo tent. There was absolutely no way I'd be leaving my tent before daylight. Thankfully falling back to sleep, I awoke early enough to get up and watch the sunrise. A thoughtful host brought me a cup of tea and I watched as the mountains around were gradually lit by the sun. The monk's chants and trumpet could be heard alongside the birds heralding a new day, my last full one in Bhutan. Over breakfast as I shared my tale of midnight purring, it reminded Wangchuk about my waking during the nights in Bumthang, "perhaps they were spirits", he said. "Bumthang is known for

Bumdrak camp.

troublesome spirits." I thought of the left knee of the demoness pinned down by Dzongs and temples and wondered if she was restless in her deep sleep. After some impressive bird spotting of giant blue and turquoise pheasant like Himalayan Monal, we began our descent. There was another summit above Bumdrak, that we didn't have the time or inclination to conquer this visit. Namgo La (the 'pass as high as the sky') was another 45 minute climb and 4100m. Prayer flags marked it and I would have loved to have seen what the view was from it. It is used as a sky burial site. This is a traditional burial where the deceased's remains are carried to the area then with a ceremony, presented to decompose exposed to the elements and eaten by scavenging animals, particularly carrion birds. As Buddhists believe that the body is an empty vessel, and living in rocky terrain where digging wasn't practical, sky burials were a way to dispose of remains in as generous a way as possible. I first learned of this practice as a child and found it quite frightening, along with being buried or burned alive. Now I think it would be an incredible honour. 🐝 The walk down was easier than the climb up, however in wet weather it would be terrifying as the pathway was, for us, dusty soil. Wet weather could turn it into a lethal slide. We passed Yoselgang temple and monastery with prayer wheels over a stream so the water turned the wheel ringing bells, I

Bumdrak Summit prayer flags.

can hear that sound now when I wish to remember it. We met a few pilgrims climbing up but otherwise we were alone, finding the pathways and enjoying the views. Ugyen Tshemo Lhakhang was a beautiful temple above Tiger's nest with a huge prayer wheel. It was closed as we passed down an almost hidden trail through larch and silver pine trees. We then arrived at a beautiful gateway leading to Zangto Pelri Lhakhang. There was a viewing platform here down to Tiger's Nest. I could have stayed here a lot longer. We found a huge bush of

flowering ivy, covered in bees, Apis mellifera alongside Apis dorsata, all feeding on the nectar. We must have looked odd filming the ivy with our backs to the most splendid view of Bhutan's most famous temple! Soon after descending from this view point, we began meeting a steady stream of tourists as our path joined the route to Tiger's Nest. 🐝 As we had descended to the temples, I was surprised by the valley we had to cross, hundreds of steps down, then hundreds more up the

Tigers nest at Bumdrak.

other side. Panting tourists on both sides. As we arrived at the site, all bags, phones and cameras had to be handed in to lockers as we visited each of the many separate temples each attached to the cliff face, apparently by the hairs of Khandroma (a female celestial being). The site has long been recognised as a holy place and Guru Rinpoche flew to the site on the back of a tigress (a manifestation of his consort Yeshe Tsogyal) to subdue the local demon. He then meditated in a cave there for three months. The primary temple was built by the governor (penlop) of Paro in 1692 Much of Tiger's nest was rebuilt after a fire destroyed it in 1998. It was reconsecrated in the presence of the king in 2005.

Bumdrak temple steps.

There was a lot to see here and magnificent views, but also a long steep walk down with a lot of tourists and their guides to fill the trail. There is a fabulous café about half an hour below, where the less able can wait whilst others climb the remaining path to Tiger's Nest. We had our lunch here whilst chatting to a large group of Bangladeshi Christians making their pilgrimage. We sped down the remaining path and were overjoyed to see Sonam waiting for us among the market stalls at the base.

24. GROSS NATIONAL HAPPINESS

T WAS MY LAST EVENING IN
IN BHUTAN. COMPLETING

the Bumdrak trek was exhilarating and Wangchuk and I had lots to share with Sonam about our adventure. It had been Wangchuk's first time too and we both thoroughly enjoyed it. It didn't take long to pack my bags. I removed anything unnecessary, or heavy and distributed it amongst the hotel staff and my lovely guides. I had a collection of heavy iron gifts, honey and my 'yathra' rug from Ura. Before our last dinner together, I sat in the large balconied living room, where a week earlier we'd shared my birthday cake. There were books and magazines around, and I stumbled on a large hard backed copy of 'An Extensive Analysis of GNH Index' from 'The centre for Bhutan Studies'. I wished I had more time to read it properly, or the luggage allowance to take home my own copy. Skimming through during the time I had, I was fascinated that a government would take so much trouble to

Last dinner and night in Bhutan.

discover the happiness of its people. Regular questionnaires were sent to residents around the kingdom asking detailed questions about education, family, careers, and their level of happiness now, in the past and how happy they would expect to be in the future. The most recent results were charted and percentages of happiness were categorised amongst regions and between male and female, old and young. A pie chart showed the percentages of contributors to unhappiness including health, education, community vitality, ecological diversity and resilience, psychological wellbeing, cultural diversity and resilience, good governance, time use and living standards. Health and community vitality being the lowest contributors to unhappiness. The largest contributors to unhappiness were education (lack of) and living standards. Perhaps the increase in global awareness through the internet and television had highlighted the living standards of those seen on the screens and lack of education on global affairs had created a shock at seeing the gap between Bhutanese ways of living and seemingly, the rest of world. The government are using these statistics to improve the lot of their people. The impact of happiness is considered with every business plan, government decision and establish an increase in happiness across the kingdom. Bhutan may not be the happiest place on the planet with 10. 4% of their population being 'unhappy' and 59% of people being 'not-yet happy'. How can we know though when no other country has even attempted to quantify our happiness? How incredible that the King and his government are taking action and defining decisions based on Gross National Happiness (GNH) rather than the rest of the world's Gross Domestic Product (GDP) measure of success. I had certainly been very happy throughout my visit and witnessed much laughter and contentment from those I met. After my return I realised just how happy the Bhutanese were, or rather, how 'un angry' they were. I noticed how in airports people were impatient, the roads aggressively driven on and the radio and television full of angry and bitter people. I couldn't watch many programs on my return, offended by the level of anger and hatred displayed by so many for such petty things. We all have so much to be grateful for, even the simple fact of being alive, it certainly beats the alternative.

25. LEAVING BHUTAN

I T WAS APPARENTLY MOST AUSPICIOUS TO PASS A DEAD BODY BEING

prepared for transportation for cremation. On our last drive together to the airport, we paused to learn why the road was filled with cars and people at such an early hour. A crowd had gathered around a farmhouse, and I could see a bright orange and yellow canopy which had been placed over the container with the body. 🐝 It was only a brief moment but after the now so familiar sigh, more of compassion than disappointment, from my guide and driver, I learned of this customary ritual and how lucky we were to start our day seeing it. 🐝 It was hard to keep the tears away as my driver of the past sixteen days, Sonam, plugged his USB into the car and turned up the volume of the selection of Bhutanese chants and songs that now had become so familiar. 🐝 'Penor', with his Hawaiian style prayers over a melodic tune brought back memories of our twisty road journeys. This Bhutan adventure has reminded me often of my trips to Hawaii, another spiritual centre. Penor's music reminded me of Israel Kamakawiwo'ole. 🐝 What was making me so sad to be leaving? This sadness was deeper than the 'end of a holiday'. Despite looking forward to returning home, I really couldn't bear the thought of never returning to this land of the Thunder Dragon. I had made so many new friends, met up with longer term 'virtual friends' and joined the community of those who've 'been to Bhutan'. 🐝 Sitting next to a Bangladeshi Christian in the café below Tiger's Nest, he told me of his groups trip to Bhutan, to learn and meditate. His recollection of their guide telling of Bhutan's three guiding principles, no anger, no greed and no ignorance. Although I'd not heard of this before, I was very aware that I'd not witnessed or sensed a moment's anger during my stay. 🐝 The roads, twisting around mountains would give many a western tourist reason to anger. If not being behind a slow moving highly decorated lorry, sleeping dogs, cows, monkeys and farmers to give a cause of irritation, then cars overtaking on blind hairpin bends, drivers trickling out in front, stopping or turning without warning would. I could see many a cause of road rage. Despite this the roads remained calm, the odd 'oh' when a manoeuvre looked a tad dangerous, giggles when any livestock blocked a lane and extreme patience when engines were turned off completely as a pot hole was filled or a digger filled a lorry with rocks or gravel. Imagine our own daily lives with anger simply removed. There was talk during my stay about how the non compulsory tipping of guides and drivers had reached an expensive peak a few years ago following over generous visitors setting a precedent. This had now calmed down, a local told me that 'any sized tip given with a good heart was much appreciated'. That was a relief knowing that no financial tip would fully show my appreciation of the tireless attention and care I received from my own guide and driver. Greed was therefore not a major issue. Often, I was reminded that those with the least give more. Never more apparent than in the temples where visitors regularly hand over large portions of their daily income to statues of Buddha or some of the other, many, deities and gurus represented around the country. As my Bangladeshi Christian reminded me, "it's all about sacrifice, be it money, time, energy or a goat". Being strictly against killing of any kind, money, food, milk and time were the currency of sacrifice in Bhutan. The more you gave, the more the gods bestowed on you in good fortune and blessings. Giving could be sat spinning a prayer wheel all day. Bhutan therefore is a country unlike any other I've visited. Preserving their landscape, nature and heritage, they have managed to escape the trappings of Western 'civilisation '. TV was introduced in the late 1990s, yet they are bemused as to why we would put chemicals on our food or on the land. Each hospital in Bhutan has its own traditional medicine department and, in the capital, Thimphu, a state of the art Traditional Medicine hospital was certainly a centre that I would love to see closer to home. Bemused that we could have lost such obviously beneficial aspects of living, the Bhutanese are pleased to have what they do and certainly don't appear eager to move West.

Many westerners are drawn to Bhutan though. I wasn't the only solo middle aged woman travelling to find peace, organic food and spiritual enlightenment. Perhaps I was the only one on a quest for bees and honey.

If I knew that I could be returning home to a place where I didn't have to explain why I want to eat organic food, use natural medicine or just the time to pause and meditate, I wouldn't be feeling so sad to leave. It may be easy to think that rural uneducated farmers maybe living in ignorance, yet I'm sure it is us ignorant to our connection to all things. We may have learned much with our education for all, yet it is the simple Yak herder or Bhutanese school child who understand truly what damage ignorance does to our earth and our nations.

As for bees, months after my return, I found a paragraph written in my notebook after asking Sonam about what the Bhutanese believe about any spiritual connection with bees and honey. I was astonished to find that I'd written; ' monks believe that the highest transformation is into a bee and then to teach other bees all you have learned, who then inform the humans.' I had also written that monks are interested in the 'song' of the honey bee, its vibrations. How extraordinary, that the messages I was longing to hear from the Bhutanese bees were there in my notebook from very early on in my trip, yet I'd been so focussed on learning who could have punished Buddha for taking honey and why, that it wasn't until months after returning that I found my lesson.

I now see that the people of Bhutan are perhaps already so spiritually advanced that they no longer need to eat honey to be healed. They are the 'bees' teaching the rest of humanity what they have learned. With the selection of honeys I brought back with me, I hope to share the learnings of Bhutan and its people. Hopefully a little bit of Bhutan will stay with me and can be shared back at home. The short song 'hello' on Sonam's usb, cut short when his hands were free to leave the steering wheel, was always greeted with a giggle and a goodbye '. I now hope the sweet Bhutanese voice will say hello to the world and inspire us to listen. Hello? Are we listening? Can we bring some of the Bhutanese wisdom to our own worlds creating more countries where happiness is measured and valued.

I hope so.

Last view of Bhutan.

Websites and reading list

Bees For Development
www.beesfordevelopment.org

Travel Counsellors
www.travelcounsellors.co.uk/mariafoxwell

Norbu Travel
www.norbubhutan.com

Drukair
www.drukair.com.bt

Opening your heart to Bhutan
www.openingyourhearttobhutan.com

Beekeeping cooperative in Bhutan
Tel: +975 03 631171
Navin Beekeeping:
+975 17 64 35 85

Bhutan Trust Fund for Environmental Conservation
www.bhutantrustfund.bt

Bailey Hill Bookshop
www.baileyhillbookshop.com

Candide
www.candidegardening.com

Northern Bee Books
www.northernbeebooks.co.uk

Recommended Reading List

Artist to Bees (1st edition)
Paula Carnell

Artist to Bees (2nd edition)
Paula Carnell

Bhutan
Lonely planet

Set free
Emma Slade

The Tibetan Book of Living and Dying
Sogyal Rinpoche

Radio Shangri-La - What I learned in Bhutan, the Happiest Kingdom on Earth
Lisa Napoli

Specialist bee book suppliers;
www.northernbeebooks.co.uk

PAULA CARNELL

Paula Carnell was born in Dorset, England and has spent much of her adult life living in Castle Cary Somerset. Forming 'Possi' in 1990 as part of the Prince's Youth Business Trust' scheme, she soon had a successful enterprise selling her original paintings on silk, and printed greeting cards of her work in over seven hundred shops across the UK and exporting to eleven countries worldwide. Opening a gallery in Castle Cary in 1995 established her as a familiar face in the town, until she 'retired' from retail in 2004 and focussed on her personal painting career.

Photograph by Janette Edmonds.

Exhibiting in London and the USA, Paula was fulfilling her dream as a globe travelling artist. Then in 2008, she began to fall ill, becoming bed and wheelchair bound with Ehlers Danlos Syndrome in 2009. The following seven years were spent on a personal quest to find meaning in life, transforming from an artist to a bee speaker. Achieving a full recovery in 2016, she is studying as a medical herbalist with the IRCH, runs her business 'Creating a Buzz about Health', working as a Global beekeeping consultant, writer and speaker. She lives in Castle Cary with her husband Greg and three sons.

ADDITIONAL BOOKS BY THE AUTHOR

THE RIME OF THE ANCIENT MARINER
Illustrated by Possi
Published by Elite Words and Images 1993

FORTY FLOWERS
Illustrated by Paula Carnell
Published by P. C. Paintings 2008

PAINTING PROFESSIONALLY ON SILK
Published in 2008

ARTIST TO BEES (1ST AND 2ND EDITIONS)
Published in 2019

You can subscribe to Paula's newsletter
and receive updates to her blog,
purchase copies of her books
and the 'Plants for Bees' Tea towel at:
www.paulacarnell.com

Apis mellifera on Bumdrak trek.

'Plants for Bees' Tea towel available at: www.paulacarnell.com

Follow Paula's weekly bee updates on the Candide App.
www.candidegardening.com

PLANT BASED MINERALS

If you are interested in learning more about the plant based minerals, follow Paula's blog above, or visit her minerals website:
www.carnellsminerals.com

Apis mellifera at high altitude on the
Bumdrak Trek 3100m.

PAULA CARNELL

Creating a buzz about health